FOAL

blue
rider
press

DEFYING REALITY

ALSO BY DAVID M. EWALT

Of Dice and Men:
The Story of Dungeons & Dragons and the People Who Play It

DEFYING REALITY

The Inside Story of the Virtual Reality Revolution

DAVID M. EWALT

BLUE RIDER PRESS

NEW YORK

blue
rider
press

An imprint of Penguin Random House LLC
375 Hudson Street
New York, New York 10014

Blue Rider Press is a registered trademark and its colophon is a trademark
of Penguin Random House LLC

Portions of this book were originally published in *Forbes* or on Forbes.com.

Library of Congress Cataloging-in-Publication Data has been applied for.

Hardcover ISBN: 9781101983713
Ebook ISBN: 9780735219779
International edition ISBN: 9780735215672

Printed in the United States of America
1   3   5   7   9   10   8   6   4   2

Book design by Daniel Lagin

For my grandfather Michael E. Spinapolice,
who bought my first computer

CONTENTS

Prologue

THE SPARK

I t's 1894, and a twenty-year-old Italian aristocrat pushes a button on his desk, causing a bell to ring on the other side of the room. He wakes his mother in the middle of the night to show her what he's created. He calls it a wireless telegraph.

It's 1927, and a twenty-one-year-old Utah farm boy transmits a live image from a camera, through the air, to a glowing screen. "That's it, folks," he announces. "We've done it—there you have electronic television."

It's 2012, and a nineteen-year-old video game fan from California fits a lightweight plastic headset over his eyes, presses a button on a computer, and is transported to another world. "I am making great progress," he tells his friends later in a post on an Internet forum. "Really excited about this."

We've already entered the age of virtual reality, though you probably haven't noticed it yet. You've almost certainly heard of VR, seen the news stories and magazine covers, read about how it's the hot new medium for 3-D movies and video games. It's possible you've tried one of the basic VR viewers that use a phone for a screen. Or maybe you've played with or own a high-end VR headset that connects to a computer.

But unless you're one of the handful of people who live on the sharpest point of the cutting edge, you probably haven't noticed that the release of those gadgets was the dawn of a whole new era. This isn't just another beat in the accelerating tempo of techno-logical progress; it's the start of a brand-new song. At the very least, it's a moment as significant as the birth of radio or televi-sion; quite possibly, it's the beginning of a fundamental change in what it means to be human.

No, seriously.

I know that sounds crazy. But this technology gives us the ability to do crazy things. A virtual reality is a computer-generated environment that you can see and hear, typically through the use of a high-tech headset, so that it appears you're actually inside the simulation. Good VR even lets the user interact with and change the environment. Now think about that: Creating a whole new world that people can inhabit used to be something only deities could do. The ancient Greeks said Gaia gave birth to the heavens, the sea, and the mountains; in the twenty-first century, an engi-neer models them on their laptop.

And think about what it means to inhabit one of these virtual worlds. You and I are bound to the physical world—we have to work with the body we have in the place where we are. But as virtual reality simulations get better, both of those limitations start to go away. Suddenly anyone can see what it's like to stand on the peak of Mount Everest. Or a person who can't walk can experience a marathon from the perspective of an Olympic champ. And if fantasy is indistinguishable from reality, why stop there?

Take a walk across Mars—hell, take a walk across Narnia. Become a dragon and fly through the clouds.

Crazy, right? I don't know what it will do to humanity when we can experience our fantasies in a manner that's indistinguishable from real life, but I do know that the invention of this technology is a pretty big deal. When we talk about VR, we're not just talking about gadgets that play 3-D video games.

———

In the interest of transparency, I should admit that I'm exactly the type of person you'd expect to get overexcited about VR. When I was a child I was obsessed with fantasy literature, movies, and games. I memorized the maps in books like *The Hobbit,* knew the name of every obscure *Star Wars* character, and spent countless hours playing Dungeons & Dragons with my friends. I loved video games too, and I was interested in computer programming. In other words, I was a stereotypical 1980s nerd.

Then I read William Gibson's *Neuromancer,* and it blew my adolescent mind like a bolt of neon lightning. In this cyberpunk novel's dystopian future, hackers transfer their consciousness directly into the Matrix, a virtual reality representation of a global computer network, and explore cyberspace the same way a team of adventurers might delve into a dungeon in one of my D&D games. Gibson's "console cowboys" navigate traps, fight powerful security programs, and escape with stolen data treasure. A world where you can fight monsters and still be a computer genius? I would have moved there if I could.

I spent my teenage years devoted to anything that had to do with virtual reality. I devoured books written by novelists like

Gibson, Bruce Sterling, Rudy Rucker, and Neal Stephenson, and used my allowance to buy a subscription to *Wired* magazine. I lost interest in D&D and started playing sci-fi role-playing games like Shadowrun and Cyberpunk 2020, where instead of a knight or a wizard, my character could be a netrunner or a decker. I even paid to see cyberpunk B movies like *The Lawnmower Man* and *Johnny Mnemonic* in theaters—multiple times. And I spent what little free time I had left on my computer, connecting to dial-up bulletin-board systems with a 2400-baud modem, imagining I had become a console cowboy. I was on the VR bandwagon and couldn't wait for the future.

But then I got burned. When I was eighteen years old, the Japanese video game company Nintendo announced the Virtual Boy, a portable video game console that could simulate immersive 3-D graphics. It looked like something out of one of my cyberpunk novels—a futuristic headset in red and black plastic. For the better part of a year before it was actually released, TV commercials and articles in *Nintendo Power* magazine raised my expectations to epic levels. At night I dreamed of video game characters jumping off TV screens and running all around me in glorious virtual reality. When it was finally released, the Virtual Boy cost nearly $200. But I was away at college, so I bought one with the money my parents gave me for textbooks.

It sucked. Oh, sweet Mario of the Mushroom Kingdom, how it sucked. The Virtual Boy had a weird red-and-black monochrome screen that gave me eyestrain and migraines. The graphics weren't immersive 3-D, they just had an illusion of depth thanks to stereoscopic trickery. And the games were nothing special—pinball, tennis, baseball, nothing that made me feel like

I'd been whisked away into another reality. Just a few minutes of play made me want to vomit—half from disgust and half from the pounding headaches.

I was crushed. This gadget was supposed to be my point of entry into a world of VR fantasy, but instead it was just an over-priced View-Master. And I wasn't alone; consumers around the world hated the Virtual Boy, making it one of the biggest flops in Nintendo's history. I tossed mine in my closet, and it sat there unused and unloved until the end of the school year, when I sold it rather than bothering to pack it up to take home for the summer.

From that point on, I was a hard-eyed skeptic when it came to the topic of virtual reality. I never stopped enjoying portrayals of virtual worlds in fiction—I think I saw *The Matrix* four times when it was in theaters. But I was instantly suspicious of any company, scientist, or engineer who claimed to have developed a viable VR product. I kept an eye on developments in virtual reality as I started my career as a technology journalist, but years ticked by without anyone making real progress. Research failed to pan out, consumer products didn't work, tech demos produced more nausea than converts. I wrote sarcastic blog posts about the few people who were even trying.

And then in 2012, I heard about a nineteen-year-old video game geek from California who had built his own virtual reality headset. His name was Palmer Luckey, and he called it the Oculus Rift. The few people who'd tried it swore it was the real deal—the kid had solved some major technical problems, they said, and his invention provided real immersion into virtual worlds.

I remained skeptical. I'd heard all this before, and besides, the teenage genius tinkering in his parents' garage was such a Silicon

Valley cliché, it had to be hype. But when Luckey's start-up began selling $300 Rift prototypes on the crowdfunding website Kickstarter, I couldn't help getting a little excited. The campaign was endorsed by a long list of people whose opinions mattered— people like John Carmack, the legendary programmer of video games like Doom, and Gabe Newell, the billionaire owner of video game software developer Valve. When the Kickstarter ended after thirty days, Oculus VR had raised more than $2.4 million—and for the first time in more than a decade, I felt like the dream of VR might actually be moving forward.

———

I didn't realize we'd entered the age of virtual reality until almost two years later. By that point, Oculus VR was already a phenomenon. A few months earlier, Facebook CEO Mark Zuckerberg had closed a deal to acquire the company for $2 billion, even though it still hadn't released a commercial product. Oculus was hosting a conference at a hotel in Los Angeles and showing off its latest Rift prototype to a group of software developers, engineers, and designers; I'd been assigned to write a cover story about Palmer Luckey for *Forbes* magazine, so I was able to get into the event and try out the new headset.

The demo took place in a small room with a slightly raised five-foot-square pad on the floor in the center. An Oculus technician told me to stand on it and explained that I was going to see a series of short scenes designed to show off the Rift's capabilities and put users into a variety of virtual spaces. As long as I didn't feel myself stepping off the pad, I could move around and explore, since the system would follow my location and orientation.

Then he handed me the headset—a beveled box of black plastic about seven inches wide, four inches tall, and three inches deep attached to a plastic head strap. The inside surface of the headset was gently curved and padded with foam where it would rest against my face, positioning my eyes directly in front of two convex lenses that would focus my vision on an internal LED screen. The device was surprisingly light and looked vaguely alien. I had the odd sense I was handling something I shouldn't touch, like an artifact from the future left behind by a careless time traveler.

The technician helped me put on the rig and adjusted it to sit on my head properly, so it felt no more cumbersome than a pair of ski goggles. He pulled headphones down over my ears, and because my vision and hearing were blocked, my other senses were heightened. I remember standing there as I waited for the demo to begin, feeling the air-conditioning blow on the back of my neck, the soft pad beneath my feet, the foam of the headset against my brow.

And then it was all gone, and I wasn't in a meeting room at a hotel in Hollywood, I was standing on the roof of a building, surrounded by skyscrapers, in the middle of a strange city. It was nighttime, and the towers around me were lit by spotlights and thousands of twinkling windows; above me, low-hanging clouds reflected the city's light with a sickly crimson glow.

I gazed across the skyline, saw it stretching to the horizon, and felt the place's size. It wasn't like looking at a picture of a city, it was like I was actually *in* the city; it filled my entire field of vision, and had real depth and weight. As I stepped forward, the scene moved with me until I was at the edge of the building; when I looked over, my stomach lurched with genuine vertigo as I gazed down at cars driving on the streets far below.

It felt so real. Sure, the computer graphics weren't perfectly realistic, but the overall effect was so convincing I forgot I was wearing a VR headset. It felt more like I was *there* than *here*—I was on that rooftop looking at that city, not in a demo room staring at a screen a few inches from my eyes.

Then the scene faded to black, and then faded up again, and I found myself in a small sitting room with two plush armchairs on either side of a table set for afternoon tea. There was a gilt-framed mirror on one wall of the room, and when I turned to it, I saw the reflection of a porcelain harlequin's mask floating in midair. When I leaned toward the mirror for a better look, the mask leaned in too; I tilted my head to one side, and the mask copied the movement. I yelped in surprise—"Oh, that's me!"

The scene changed again, taking me on a series of short visits to a half-dozen virtual worlds. I stood on the surface of a rocky planet with three moons in the sky and came face-to-face with a friendly green-skinned alien; it seemed so real that when it raised a hand to wave hello, I automatically smiled and waved back at it. I shrunk down to microscopic size and found myself facing a giant, truck-sized mite; basketball-sized pieces of dust and pollen floated through the air, and I walked around them, examining them from all sides. I visited cartoon animals grazing in a pasture. I squeezed into the claustrophobic control room of a dimly lit submarine.

Finally I found myself in an empty corridor with high cathedral ceilings and floor-to-ceiling windows along one wall. As I looked around the room and tried to figure out where I was, the simulation's audio kicked in, and I heard a low thump off in the distance, around a corner. Then another, this time closer. Another, closer still, seemed to shake the entire room. Then something stepped around the corner. A massive scaly head, a huge muscled leg, a long powerful tail. A *Tyrannosaurus rex*, king of the dinosaurs, as big as a bus, coming right toward me.

The parts of my brain responsible for higher reasoning knew this was just a simulation, but the simulation was so realistic it had more primitive parts of my brain utterly convinced. At the sight and sound of a massive predator approaching, some organ that has protected land-based mammals since the Triassic period took over my body, and I learned for the first time how my body handles the fight-or-flight response.

I froze in fear. I felt the hairs on the back of my neck stand up straight and my heart start to pound in my chest. As the T. rex lumbered toward me, I broke out in a cold sweat. When it stopped about thirty feet away and roared, I could swear I felt its hot breath. It came closer, close enough that I could reach out and touch it, but I still didn't move an inch. The T. rex bent down, sniffed me, and turned its head to stare at me with a hubcap-sized yellow eye. Then it straightened up, took a step forward, and walked on. My body—still out of my conscious control—finally unfroze when I ducked to let its swinging tail flick over my head.

———

The demo ended there, but after I lifted the Rift off my head, I took my first steps into another new world. In this reality, VR was fact, not fiction. After decades of false starts and failures, the technology was finally good enough to go mainstream. Unlike so many other VR demos, this one left me feeling excited, not disappointed; perhaps more important, I didn't feel like I needed to vomit. The Rift was the real deal. Virtual reality for the masses was here.

I sat by the hotel's outdoor pool and tried to think about the experience like a skeptic, not a fan. I considered that the headset was kind of bulky and made my face feel a little hot after a while. The graphics were comparable to those of most modern video games, but far from perfect—even a high-definition screen appears blocky and pixelated when it's two inches in front of your eyes. And though I hadn't seen the computer that was connected to the Rift, I knew it must have been an absolute powerhouse, probably costing thousands of dollars to build. The Rift was a high-end gadget for people with lots of disposable income, not the kind of product the average family would buy and play with in their living room.

But prices would drop and the quality would increase over time. The important fact was, it worked. I'd seen it and now I believed it: VR was going to be a big deal. This was the birth of an industry that will be as big as telephones or television, but will have an even more profound effect on our lives. Forget talking to a faraway relative on your cell phone; imagine interacting with a live virtual person that matches that individual's appearance, mannerisms, and movements, so it feels like you're both in the same room. And how can two-dimensional TV and movies compete with immersive VR? Virtual reality entertainment will introduce new avenues of expression and deliver unprecedented levels of realism in art. I always thought *Jaws* was an amazing movie. Now I was gonna need a virtual boat.

The more I thought about it, the more excited I got. There are applications for this technology in every business and every home. Imagine educational software that lets students shrink to the size of an atom and study chemical reactions up close, then fly into space and witness the first steps on the moon. Virtual clothing stores where customers can see how an outfit looks on a perfect simulation of their own body. Design tools that let engineers sit inside and drive a new car prototype before it's been built. Real estate open houses where buyers tour a new home without leaving their own home. Or wars fought entirely by robots, each one controlled by a soldier wearing a VR rig in a command center thousands of miles away.

I was getting ahead of myself. At that moment, I needed to stick to the facts and take careful notes about the Rift demo. I took out a notebook and started writing—what the hardware looked like, how it felt when I put it on my head, what the graphics looked like, and the plot of each short scene. But I kept having to stop because I wasn't sure how to describe what had happened. Was I merely recording observations of a piece of media, like a movie reviewer recounting the plot of a film? Or was I making a memoir of an event I'd personally experienced?

As a reporter who writes about video games, I was used to describing participatory entertainment. Even though we as players experience the action through a fictional protagonist, we say, "I defeated Bowser and saved the princess," not "I pressed a button that caused a collision detection function to trigger an animation indicating the game was over." But the Rift had pushed the idea that gamers become their character to a whole new level. There was no hero character acting as an intermediary between me and the creatures I met in VR, and no TV screen or monitor physically separating me from their world. My whole life I'd been playing games that required me to ignore what I could see and feel in the real world and imagine I was somewhere else. But the Rift had completely flipped the script. VR doesn't need you to ignore your senses—it needs you to embrace them. If anything, the challenge is to remember that what you're experiencing isn't real.

For all practical purposes, I had visited a city of skyscrapers. I had waved at an alien. A dinosaur had roared at me, and my body reacted like it was actually there. What did that mean? It surely didn't feel like a game. And this was a first-generation product, comparatively primitive. What would happen when VR inevitably improved to a point where its simulations were indistinguishable from the real world? If I visit a virtual Paris that looks, sounds, smells, and feels exactly like Paris, then how is my experience any different from that of a person who physically travels there?

I'd been dreaming about virtual reality since I was a kid, but I wasn't prepared for the reality of VR. My expectations were based on limited experience. I'd interacted with computers through graphical user interfaces that employ the metaphor of a desktop with files and folders, so that's the way I thought of VR—an amazing way to access information, but basically just Microsoft Windows in immersive, realistic 3-D. It never occurred to me that a VR experience might trick my brain into thinking it was real, because in the thousands of hours I'd sat in front of a screen, I'd never believed my computer's desktop was an actual physical table.

But there I was, my brow still damp with "oh no, a dinosaur is going to eat me" sweat, drying off in the California sun. And that's when I started thinking that VR isn't just going to be a new tool for business or a new medium for 3-D movies and video games.

It's going to change what it means to be human. This technology could allow us to escape the bodies we're born in and the geographies that confine us. It could allow us to experience the impossible, to do the unthinkable. What happens to humanity when you can experience at the push of a button what it's like to kill a person, or to have sex with them? Will we still feel the need to explore when the bottom of the ocean or the surface of Mars is just a click away?

The more I thought about virtual reality, the more I felt like one of the primitive apes at the beginning of Stanley Kubrick's *2001: A Space Odyssey*—the monolith has appeared in our midst, and all I could think to do was throw handfuls of dirt at it. I wanted to know how VR got here, how it worked, and where it was going; I wanted to understand how it might change our culture, our communication, and our consciousness.

————

So this book is an attempt to reach out and touch the mono-lith. It chronicles some of the things I learned and experienced in just over two years during the most explosive period of growth in the history of virtual reality—a time when we went from zero commercially viable products to millions of units sold by some of the most important companies in the tech industry. It documents the origins of virtual reality, from its birth in military labs, through the booms and busts that nearly doomed the technology, to the twenty-first-century inventors who learned from history and finally brought VR to the masses. And it also looks forward, examining how this remarkable tool might change different industries, pastimes, and the way we all relate to one another.

This book has also been an excuse for me to indulge those old childhood dreams of being a console cowboy, and try to get more people to join me in the virtual world. By writing it, I'm hoping to share all the cool stuff that's happening right now, but I also want to get people excited. I want more hands making better hardware, smarter interfaces, and lots of great VR content. And I also want more people to get involved in the conversation. Right now, a small group of people are building a place where we're all going to spend a lot of time in the future. They're amazing people, but that's too big a job to leave in the hands of a lucky few. We need everyone involved: women and men, adults and children, rich and poor, able and disabled. We need computer scientists collaborating with philosophers, ethicists, and artists. We've got the chance to build a new world, so let's make it for everyone.

The more I've learned about VR, the more I've become aware of a gulf that's dividing our present and future. Most of us stand on the near side and wonder what's across the water. Maybe they haven't had the time, money, or opportunity to experience virtual reality. Or maybe they've tried it and still remain skeptical. But as the tech gets cheaper and better, every day more people will come across the water. First by the thousands, soon by the millions, and before you know it, virtual reality will be a part of daily life for billions of people.

I hope this book will be a message in a bottle encouraging you to take the journey.

Chapter 1

PYGMALION'S
SPECTACLES

There is a unicorn in a cavern under an ancient forest in France. I've seen it myself, even though it wasn't really there; I've stood in front of it, even though its home is a place where no one can go.

The unicorn is painted on the wall of Lascaux Cave, a subterranean complex in the Vézère Valley, about two hours' drive east of Bordeaux. Seventeen thousand years ago, some of the first modern humans in Europe decorated the caverns with hundreds of images of animals, symbols, and abstract shapes. In 1940, it was rediscovered by a group of teenagers and became one of the most famous archaeological sites in the world.

For decades following the discovery, countless tourists flocked to Lascaux. Henri Breuil, a French priest and archaeologist who was one of the first researchers to study the caves, called it the Sistine Chapel of prehistory; after artist Pablo Picasso saw the paintings, he lamented that his entire generation of revolutionary artists had "invented nothing new." But the ancient drawings were fragile, and the presence of so many visitors took a toll. In

1963, the French government banned the public and locked the doors.

Fortunately, technology allows a modern tourist to sneak inside. Lascaux comes to life in 360-degree videos and virtual reality tours; I'd seen photos of the caves and studied their paintings in college, but it took seeing that unicorn in VR to really understand the place. Even though the French government has built detailed replicas of the caves for people to visit, the virtual version makes for a better trip—it's appropriately claustrophobic and free from the distracting presence of other tourists.

It's also a particularly fitting way to visit, because Lascaux may be the oldest example in human history of an attempt to create a virtual world. You'd never know it from looking at photographs, but when you stand inside the caves—virtually or in person—the intention is clear. Just like VR headsets block the real world from view, the caverns separate a visitor from the forest above. Instead of drawing with pixels on a screen, the cave artists used pigment on rock walls. Lascaux's creators used the topography of the caverns to create immersion: a chamber decorated with sketches of wild horses isn't just a jumble of drawings, but a herd that surrounds the viewer. They used perspective tricks to make the illustrations seem three-dimensional: the body of an ox is presented in profile, but its head is turned to face the viewer. And they used the shape of the rock to give their art depth and form; for example, a twist in a wall makes a deer appear to turn away as it dashes around a corner.

No one knows why the people who created the caves went to so much trouble. Perhaps the goal was to enthrall the viewer in order to teach them something, like an Upper Paleolithic version of a flashy how-to video on YouTube: As writer and technology expert Howard Rheingold suggests, "subterranean cyberspaces" like Lascaux may have served to "imprint information on the minds of the first technologists." Or maybe the caves were immersive entertainment, an attempt to tell a story in the most realistic way possible, to convey the excitement of hunting without the risk of getting gored or trampled.

In the millennia after the cavemen left their caves, increasingly advanced civilizations created increasingly sophisticated immersive entertainment. Around 2,500 years ago, the ancient Greeks built their own VR viewers out of wood, stone, and marble—the word *theater* comes from the Greek verb *theasthai*, or literally, "to view." These technological marvels exploited then-cutting-edge science in order to trick audiences, including acoustics (semicircular amphitheaters enhanced sound waves, allowing large audiences to hear voices at a distance) and mechanics (a wooden crane called a mechane lifted actors into the air and made them appear to be flying; it's the device that made possible the deus ex machina). Elaborate costumes, handheld props, and painted scenery panels helped complete the illusion, transforming the stage into a fantastic environment.

Other cultures created virtual worlds through visual arts like sculpture and painting. Qin Shi Huang, the first emperor of China, started building a city-sized mausoleum around 246 BC, filling it with terra-cotta sculptures of his soldiers, pets, musicians, and palace officials—a kind of simulated palace for him to occupy in the afterlife. The Romans used frescoes to transform their homes from dingy concrete into more pleasant realities, covering entire walls with illusionistic images. Buildings in the city of Pompeii, preserved after the eruption of Mount Vesuvius in AD 79, reveal interior rooms painted so that they appear to have balconies and windows that open up onto lush imaginary landscapes.

Chinese artists created immersive art using paintings on long scrolls of paper. Gu Hongzhong's eleven-foot-wide *The Night Revels of Han Xizai* is a record of an actual night in the life of a tenth-century government minister. According to legend, it was created so emperor Li Yu could witness Han Xizai's debauchery, including listening to music and watching women dancing. In the twelfth century, the artist Zhang Zeduan created a ten-by-seventeen-foot depiction of the city of Kaifeng, *Along the River During the Qingming Festival,* which includes accurate images of buildings, animals, and more than eight hundred people. By virtue of their size, real-life subjects, and narrative content, these panoramic paintings amount to another kind of virtual reality that transports the viewer back to dynastic China.

In 1787, the Irish-born artist Robert Barker took that idea a huge step further. Inspired by the sweeping 360-degree view from Calton Hill, a spot in central Edinburgh, he painted a massive seventy-foot-wide watercolor landscape of the city and hung it on the inside of a cylindrical surface. Viewers stood on a platform in the middle, surrounded by the painting, so that it simulated the experience of actually being in Scotland. He patented the technique as the Panorama, coining the now common term from the Greek words *pan* and *horama*—"all that which is seen."

After first exhibiting the work in Edinburgh and Glasgow, Barker took it to London. Reviews were good, but audiences were limited, so Barker created another, bigger panorama, this time of the London skyline. English viewers liked that better, and the artist earned a small fortune selling tickets to see it. In 1793, Barker went even bigger, constructing a building in Leicester Square that could display two immense panoramas, one of them stretching floor to ceiling in a three-story-tall rotunda. The new venue became an immensely popular attraction and showed a variety of scenes, including foreign cities and historical military battles. By the early nineteenth century, panoramas were a worldwide phenomenon, with similar sites constructed in cities across the globe. The immersive experiences were even touted as a high-tech alternative to actual travel.

In 1822, the French artists Charles-Marie Bouton and Louis Daguerre combined the panorama's massive imagery with elements of theater to create an attraction they called the Diorama—scenes made of multiple paintings on linen curtains over on a stage, constructed in such a way that when bright lights were shone on different spots, the images would appear to move and change. The shows were a big hit and were also copied around the world; later versions added sound effects, stage props, and even live actors to enhance the illusion. (Fueled in part by his success with the Diorama, a few decades later Daguerre pioneered another way to render realistic images—the early photographic process known as the daguerreotype.)

In a way, all art strives to create a virtual reality, to transport an audience and immerse them in a subject or story. But theater and paintings are only distant relatives of today's high-tech 3-D headsets—the direct ancestors of modern VR came out of the study of optics.

Humans are able to view the world in three dimensions because our brains combine a number of different visual cues to
create the perception of depth. Monocular cues require only one
eye and provide information largely based on context, including
the relative size of objects (a car that appears to be tiny is probably
farther away than one that appears to be huge), occlusion (if one
car blocks the view of another car, you can tell which one is
closer), and motion parallax (when you're driving in a car, objects
on the side of the road move out of your field of view faster than
objects that are far away).

Then there are binocular cues, which work because we've got
two eyes spaced apart laterally on the front of our head. Each eye
views the world from a slightly different perspective. You can see
this by closing your right eye and looking through your left, then
closing your left eye and looking through your right. Do that rap-
idly a few times and you'll notice a difference in the apparent position of objects. That's called parallax, and when your brain
notices it, it can compare and contrast the images to calculate
distance. The result is one of our most powerful binocular cues,
stereopsis, and it's the foundation of the simulated depth in virtual reality and 3-D movies. (Another binocular cue, convergence,
is drawn from the inward movement of your eyes when they look
at something up close; to test it, touch your finger to your nose and
focus on it, and then hold it at arm's length and look again. That
movement you felt helps your brain figure out the relative distance
of nearby objects, and it's actually the cause of some problems in
modern VR, since your distance to the screen never changes.)

The English scientist Sir Charles Wheatstone first identified
stereopsis in an 1838 paper called "Contributions to the Physiol-

ogy of Vision," which described how the eyes capture two perspectives that the brain fuses into one three-dimensional image.
It also outlined a device he'd invented that could "enable any person to observe all the phenomena in question with the greatest
ease and certainty"—an instrument he called a stereoscope.

It had a simple design: two mirrors attached at a common
edge to form a right angle, and two wood panels, one facing each
mirror at a 45-degree angle (viewed from above, it would look like
the letters I V I). The viewer would position himself with his nose
against the outside corner joint of the mirror so that his left eye
saw the reflection of the left-hand panel, and his right eye saw the
right-hand one. Each panel would hold a slightly different illustration of an object or scene. The viewer would look straight ahead
and see two perspectives of what appeared to be the same object,
and his brain would combine them into a single three-dimensional
image.

In his paper, Wheatstone focused on how a few simple line
drawings—cubes, cones, and pyramids—could be made to appear
as if they were three-dimensional objects. But he also understood
that with better graphics, stereoscopic devices might present convincing virtual realities. "Careful attention would enable an artist
to draw and paint the two component pictures, so as to present to
the mind of the observer . . . perfect identity with the object represented," he wrote. "Flowers, crystals, busts, vases, instruments
of various kinds . . . might thus be represented so as not to be
distinguished by sight from the real objects themselves."

After Wheatstone demonstrated his invention at an annual
meeting of the British Association for the Advancement of Science, one of his rivals picked up the idea and ran with it. The

Scottish physicist Sir David Brewster was already well known for inventing the kaleidoscope, an optical toy that had become phenomenally popular, and he spotted a similar opportunity in 3-D imaging.

In 1849, Brewster patented his own stereoscopic viewer that used lenses instead of mirrors. Users peered through two prisms at a piece of cardboard printed with two side-by-side eye images. Since the prisms bent light, the left image appeared on top of the right image in the center of the card, creating a single stereoscopic picture. The prisms also magnified the images, which meant they could be more detailed and printed on smaller pieces of paper. The device could even view stereo photographs on glass slides produced via Louis Daguerre's daguerreotype process. And since the whole rig fit into a handheld wooden box the size of a pair of binoculars, it was easy to set up, use, and carry.

Brewster's stereoscope made its debut at the Great Exhibition of the Works of Industry of All Nations—the first world's fair, held in 1851 at Hyde Park in London—where Queen Victoria tried the device. Her Majesty was amused, and her approval kicked off one of the biggest fads of the era, driving huge demand for stereoscope viewers and imagery. Over 250,000 image cards (or stereographs) were sold in France and England in just three months after the royal viewing, depicting everything from landscapes to photographs of newsworthy events to formal portraits of world leaders. Within five years, Brewster had sold around half a million viewers, and imitators sold countless more competing products, including deluxe units mounted in mahogany cabinets with polished brass hardware. A high-end stereoscope was a must-have item in any respectable Victorian parlor.

Across the Atlantic in the former colonies, one of the United States' most famous thinkers saw a need for a populist alternative. In 1861, the author, poet, and physician Oliver Wendell Holmes Sr. created what became known as the American stereoscope—a simple inexpensive viewer for the masses. His design was minimalist: two lenses in a wooden frame, an eye hood made out of pasteboard, and a small bracket for holding cards, all attached to a central spine with a short handle on the bottom.

"I felt sure this was decidedly better than the boxes commonly sold . . . and could be made much cheaper than the old-fashioned contrivances," Holmes wrote later in *The Philadelphia Photographer.* He also believed that money could be made from the device, but he wasn't looking for profit, so he tried to give the design away to anyone who would manufacture it.

"I showed it to one or two dealers in Boston, offering them the right to make all they could by manufacturing the pattern, asking them nothing," Holmes wrote. "They looked at the homely mechanism as a bachelor looks on the basket left at his door, with an unendorsed infant crying in it."

But hobbyists saw the value of the device right away. Holmes's friend Joseph Bates was a wealthy merchant and amateur photographer, and after he saw the design, he decided to build one of the viewers for his own use. Bates improved on Holmes's prototype in a few key ways, including the addition of a focus adjustment slide and wire card holders, and after he built one for himself, he made and sold a few more to fellow stereoscopy enthusiasts. From there it caught on quickly, and by the beginning of the twentieth century, the Holmes-Bates stereoscope was as ubiquitous in American homes as the television would be a hundred years later.

Holmes probably never anticipated how much demand there would be for his humble skeletal stereoscope. But he certainly understood the potential for virtual worlds to disrupt technology and society. "Form is henceforth divorced from matter," he wrote in an essay for *The Atlantic* magazine. "In fact, matter as a visible object is of no great use any longer, except as the mould on which form is shaped. Give us a few negatives of a thing worth seeing, taken from different points of view, and that is all we want of it . . . We are looking into stereoscopes as pretty toys, and wondering over the photograph as a charming novelty; but before another generation has passed away, it will be recognized [as beginning] a new epoch in the history of human progress."

———

I grew up with an antique Holmes stereoscope in my house, though I didn't know what it was at the time. My father kept it on top of a tall bookshelf with some other old stuff I wasn't supposed to touch, so of course sometimes when I was bored I'd stand on the couch and take it down. I remember being equal parts intrigued by the strange device and slightly grossed out by its age—it was always a little dusty and had an old, stale smell. I'd hold it up to my eyes, look through the lenses, and pretend it was something interesting, like a submarine periscope or an eye-mounted laser cannon. It was usually entertaining for a minute or two at most. Years later, after I started researching this book, I realized what the stereoscope was, and the next time I visited my parents, I made a beeline for its place on the shelf.

It had the same scent I remembered, a combination of rust and old wood varnish. The aluminum eye hood was pitted with oxidation, and the crossbar that held the stereograph cards was wrapped with a piece of yellowed tape. A faint design stamped into the metal gave a hint to the object's origin story—an angel flying past an ornate Art Nouveau building with a barrel-vaulted roof, and a short line of text in French: EXPOSITION UNIVERSELLE INTERNATIONALE. Our family stereoscope was a souvenir of the 1900 world's fair in Paris.

I'll never know how the stereoscope made its way from the halls of the Grand Palais des Champs-Élysées to my great-grandparents' house in Oregon's Willamette Valley. But I do know that the product is a Perfecscope, a brand introduced in 1895 by a Vermont firm called H. C. White. Legal information etched on the bottom of the viewer reveals a global demand for H. C. White's products and hints at the size of the stereoscopy fad—patents had been registered in the USA, Canada, France, Germany, Great Britain, Austria, and Belgium. When it was new, the Perfecscope was a relatively high-grade product, lined with red velvet along the edges of the engraved hood. H. C. White advertised the model as "the crowning triumph of stereoscopic invention, combining beauty, utility, strength and elegance." It sold in stores for about forty cents.

My dad kept a collection of old stereograph cards in a .30-caliber ammunition can on the bookshelf, so I pulled one out at random, fit it into the scope, and peered through the lenses. After few adjustments, a sepia-toned photograph of an unfamiliar city came into focus. It appeared to be three-dimensional, but only in the most basic sense; buildings in the foreground appeared closer than those in the back. The stereoscope created depth, but no immersion. To my modern eyes, the technology was primitive and unimpressive, not much better than looking at a landscape painting in a museum. I couldn't understand how Holmes thought stereoscopy could replace actual tourism, much less usher in a "new epoch in the history of human progress."

And then I realized what I was looking at. City streets intersecting at a 45-degree angle, with a plaza in the middle and a tree-filled park off to one side? This was Madison Square, the intersection of Fifth Avenue and Broadway at Twenty-Third Street in Manhattan. I knew this place; I'd worked in this neighborhood, walked these streets a thousand times. Suddenly I forgot my skepticism and was immersed in the illusion, almost as if I was standing on the roof of the Flatiron Building in 1905, not in my parents' basement over a century later.

Instinctively I fell into a little game New Yorkers sometimes play when they're walking down a familiar avenue, picking out storefronts and trying to remember their provenance: "When I moved here that was a bar. Then it was a shoe boutique, and now it's a frozen yogurt shop." Just like that, only in reverse. An elegant Victorian hotel? That will be an office tower in a hundred years, and the ground floor will be the 40/40 Club, a sports bar owned by Jay-Z. A corner building topped with a billboard that says *Continental Cigars Ten Cents* will be a showroom full of expensive kitchen and bathroom tiles. A man on the street headed toward the park really caught my interest—one day in the future I will trace his footsteps as I walk to Shake Shack to stand in line for a double cheeseburger.

I was lost in the image for a long time, imagining what it must have been like to dodge horses speeding down Broadway instead of taxicabs. I realized the "primitive" stereoscope had grabbed my attention just as effectively as a full-blown VR simulation. To a viewer back in 1900, Holmes's device must have been mind-blowing.

After a while, I put the stereograph away and pulled another from the ammunition can. A label printed on the card identified it as *Lakes of Killarney, Ireland,* copyright 1902 by the Keystone View Company—a Meadville, Pennsylvania, publisher that hired photographers around the world to capture local points of interest, and eventually sold more than twenty thousand different stereographic images. This one, a shot of a tranquil lake in a mountain valley, had a caption on the back that included a you-are-there anecdote so a virtual tourist might enjoy some of the same Irish wit as a real-world visitor: "An Irish writer tells us that the Garden of Eden was surely at Killarney. Here Adam and Eve spoke the Gaelic . . . but someone retorts that it was for speaking Gaelic that they were put out, and deserved to be."

Other stereograph cards in the collection depicted slice-of-life scenes that seemed to be set up and staged for cuteness (*Me an' Billy,* published 1899, a photo of a grumpy-looking toddler sitting next to a disinterested goat) or domestic comedy (*Trials of Bachelorhood,* circa 1897, shows a man with a white shirt on his lap trying to thread a needle, while two women peek around a dressing screen behind him, smirking and laughing). Several of the cards dug deeper into misogynistic humor: *Married Life as he found it—Words without music* shows a man sitting in a chair, leaning back as if exhausted, clutching his head with one hand and rolling his eyes so hard his pupils are barely visible. Behind him, a woman leans on a table, a stern look on her face, seemingly in the middle of a never-ending lecture.

A few of the cards even captured real-life moments in world history, a kind of primitive photojournalism. I lingered for a while on *President Roosevelt Seeing the Sights at the Jamestown Exposition, Opening Day, April 26 1907.* The photo doesn't capture any of the sights that might have attracted Theodore Roosevelt to a world's fair in Norfolk, Virginia—it's just a bunch of men standing around wearing dark overcoats and top hats. But I got a genuine thrill viewing a three-dimensional image of the twenty-sixth president; it kind of felt like I was in the room with him, and it gave me a better sense of his hale charisma than any two-dimensional picture could.

As I was focusing in on the last of the cards—a photo of a British army camp from the Boer War in South Africa—my six-year-old niece wandered into the room and asked me what I was doing. "I'm looking at some old pictures," I said, holding up the Perfecscope. "If you look at them through this, it makes them seem real."

She eyed the ancient stereoscope and gave me a skeptical look. When I reached over to hold the viewer up to her eyes, she hesitated for a second and then cautiously leaned forward to peer through the lenses. "Whoa, it's cool!" she exclaimed. "It looks alive. It's like magic!"

———

As devices like the Holmes scope cast their spell over nineteenth-century viewers, inventors everywhere scrambled to capitalize on the phenomenon. Most of them focused on the approach pioneered by Wheatstone and Brewster, and built hand-held or tabletop viewers packed with lenses or mirrors. But some entrepreneurs weren't satisfied with gadgets limited to a solo user. They wanted 3-D they could take public—something for which they could sell tickets to see in a theater.

That was not a simple proposition. Stereopsis works because binocular vision captures two slightly different views of the world, and the brain combines those images into one three-dimensional picture. You can simulate that for a single viewer if you use mirrors or lenses to show each of their eyes a different image. But how do you do that with an audience scattered all over an auditorium? There's not a prism on Earth that can scatter light so perfectly as to hit multiple viewers in one eye but not the other. And what about when spectators look around or shift in their seats? Even a twenty-first-century computer vision system would have a hard time tracking and targeting the position of every eyeball in a theater.

The solution was to flip the stereoscope model on its head. Instead of showing a viewer two separate images that combine into one, show them one combined image that separates into two. This kind of image, called an anaglyph stereogram, is usually composed of opposing hues—for instance, the left-eye image in red and the right-eye image in cyan—and then viewed through a pair of glasses with tinted lenses. If the right eye looks through red glass, it can't see the left-eye image; when the left eye looks through cyan glass, it can't see the right-eye image. The brain superimposes the two to create a single image that appears to be three-dimensional.

The first anaglyphs, created around 1853 by German inventor Wilhelm Rollmann, were little more than colored lines drawn on paper. But in 1858 the French physicist Joseph-Charles d'Almeida realized he could use two lanterns—one with a red filter and one with green—to shine colored light through images painted on glass slides, and they would combine to project an anaglyphic picture onto the screen of a theater. All an audience had to do was wear colored glasses and they could see a three-dimensional image, no matter where they were sitting.

D'Almeida was an academic, not a showman, so the idea didn't really go public until the 1890s, when it was picked up by French optician Alfred Molteni, the owner of a company that was one of the world's leading manufacturers of theater lighting. Molteni simplified d'Almeida's system by building a single projector with two built-in colored lenses; he called it a biunial magic lantern, and his 3-D slide shows became a popular amusement for Parisian audiences.

Molteni's productions were a big moment in humanity's long hunt for immersive entertainment. By combining the techniques of stereography with the power of theater, his magic lantern shows created a compelling artificial reality. But as the twentieth century approached, an even more convincing technology was already on the horizon.

———

It came on like a freight train—literally, in some cases. In 1895, French brothers Auguste and Louis Lumière wowed audiences with their Cinématographe, a camera that photographed a series of images on a strip of perforated film and could also project these "motion pictures" onto a theater screen. The first public showings included a series of under-one-minute films depicting dubiously interesting scenes like workers leaving a factory, a man trying to mount a horse, and parents feeding a baby. But audiences were captivated anyway. The most famous Lumière short, *Arrival of a Train at La Ciotat,* consisted of a single fifty-second shot of a steam locomotive rolling toward the viewer as it pulled into a station, and it reportedly (and perhaps apocryphally) caused members of the audience to flee from the theater, afraid that they were about to be run over.

Audiences that once lined up for panoramas, dioramas, or magic lantern slide shows now flocked to the moving pictures. When the world's first purpose-built movie theater opened for business in 1896, demand was so great the operators kept the doors open thirteen hours a day and still continuously packed the theater. In its first year alone, the seventy-two-seat Edisonia Hall welcomed 200,000 visitors.

If stereoscopy had launched a worldwide fad, moving pictures sparked a frenzy. By 1908, there were more than 8,000 movie houses across the United States. A 1910 article in the magazine *World's Work* counted 12,000 theaters, and estimated that 5 million Americans attended the "picture shows" daily—about 5 percent of the US population at the time, or one out of every twenty people. "Five-cent theatres abound on every hand," the magazine reported, and "squads of police are necessary in many places to keep in line the expectant throngs awaiting their turn to enter the inner glories."

Inside the grand picture houses and nickelodeon storefront theaters, those film audiences experienced a kind of primordial virtual reality. Early moving pictures might have been monochrome and silent, but the photographic images and lifelike motion were real enough to transport viewers into the world of movies. Audiences were whisked out of their seats and away to ancient Egypt, or to the surface of the moon, or into the middle of a gunfight in the Wild West. When an outlaw cowboy took aim and fired his pistol directly at the camera at the end of *The Great Train Robbery,* theatergoers screamed in terror, because at some level, they'd forgotten it was only a movie.

Of course moving pictures were still just flat two-dimensional images. The directing style of the time was flat too, with most films staged like a scene from a play—a single stationary camera pointed at actors lined up in front of a painted backdrop. So some filmmakers tried to make movies more immersive by using the techniques of stereoscopy. In 1915, *The Great Train Robbery* director Edwin S. Porter showed off a short anaglyphic 3-D movie at New York City's Astor Theatre; the audience wore red and green glasses and watched a few minutes of footage that included dancing girls and shots of Niagara Falls. According to a review in the *New York Dramatic Mirror,* the audience was "frequently moved to applause" and regarded the scenes as "forerunners of a new era in motion picture realism." But shortly after the screening, Porter quit directing and never improved or shared his technique.

It took a few years for other filmmakers to develop their own anaglyphic technology, but in September 1922, a feature called *The Power of Love* became the first 3-D film commercially released in theaters. The movie was shot with two cameras and presented on screens using two projectors; critics gave it positive reviews, but the setup may have been too complicated, since it played only a handful of times in Los Angeles and Manhattan. Other 3-D films failed for similar reasons. In December 1922, the Selwyn Theatre in New York City debuted engineer Laurens

Hammond's Teleview system, which required not only special projectors but mechanized viewing devices affixed to each seat in the theater. The two projectors were set up so that one frame from the left projector would show on-screen, and then one frame from the right, forty-eight times per second. As audience members watched through the viewers, a synchronized spinning shutter blocked and unblocked their vision so that each eye would see only the frames from a corresponding projector. This technique, known as the alternate-frame sequencing method, was ahead of its time—sixty years later it would power 3-D video games and amusement park attractions. But in 1922, Teleview was too expensive, complicated, and uncomfortable to be practical, so the Selwyn was the only theater to ever try out the system. (Hammond did okay, though—a decade later he invented the Hammond electric organ.)

Three-dimensional films remained an experimental novelty into the 1930s, but the idea of totally immersive entertainment had entered the public consciousness, and people had begun to imagine future applications of 3-D technology. In 1935, the pulp science fiction magazine *Wonder Stories* published a story called "Pygmalion's Spectacles," written by American author Stanley G. Weinbaum, that describes a system that looks and functions very much like what we now know as virtual reality.

In the story, businessman Dan Burke meets an eccentric professor who has invented a headset that can make a movie "very real indeed . . . a movie that gives one sight and sound . . . so that you are in the story, you speak to the shadows, and the shadows reply, and instead of being on a screen, the story is all about you, and you are in it." When Burke tries the invention—a device "vaguely reminiscent of a gas mask," with goggles and a rubber mouthpiece—he is transported from a New York City hotel room to an unearthly, beautiful forest, a place called Paracosma, or "Land Beyond the World," where he meets and falls in love with an elfin woman named Galatea. After he spends what feels like days in the sylvan paradise, it fades away to reveal that he's spent five hours sitting in a chair watching a prerecorded first-person movie. Later, the scientist explains how he pulled off the illusion:

> "The trees were club-mosses enlarged by a lens . . . All was trick photography, but stereoscopic, as I told you—three dimensional. The fruits were rubber; the house is a summer building on our campus—Northern University. And the voice was mine; you didn't speak at all, except your name at the first, and I left a blank for that. I played your part, you see; I went around with the photographic apparatus strapped on my head, to keep the viewpoint always that of the observer. See?" He grinned wryly. "Luckily I'm rather short, or you'd have seemed a giant."

> "Wait a minute!" said Dan, his mind whirling. "You say you played my part. Then Galatea—is *she* real too?"

> "Tea's real enough," said the Professor. "My niece, a senior at Northern, and likes dramatics. She helped me out with the thing. Why? Want to meet her?"

Weinbaum wasn't just the first to describe a virtual reality system, he was the first to illustrate how its artificial world could be more compelling than the real one—that people might get lost in the fantasy, lose track of time and place, even fall in love with a virtual character. And that kind of narrative power was too tempting to be abandoned. In December 1935, Hollywood studio

Metro-Goldwyn-Mayer released *Audioscopiks,* an anaglyphic 3-D movie viewed using cardboard spectacles with red and green lenses.

"Seeking a novelty to charm its fickle audiences, the company has revived the stereoscopic film," an article in *The New York Times* reported. "Metro appears to be fairly enthusiastic about its possibilities . . . the company is having 3,000,000 of the pasteboard 'eye glasses' made up at a cost, confidentially, of $3.25 per thousand."

The eight-minute film was little more than a demo reel for 3-D technology. It starts with a brief explanation of how stereoscopy works and what to expect from the movie ("You've heard talking pictures and you've seen pictures in color," a narrator intones. "Now for the first time we combine sound and color with third-dimensional pictures!"), before instructing the audience how to put on their glasses and launching into a series of short scenes where objects seem to leap out of the screen toward the viewer—a ladder poking through a window, a baseball player throwing a pitch, a man spraying water from a seltzer bottle. Critics were impressed, including the *New York Times* reporter who wrote in his article that "if there had been any women present, unquestionably there would have been screams when a magician conjured a white mouse onto the tip of his wand and poked it out, seemingly within arm's length of the innocent bystanders."

Audioscopiks worked better than the 3-D experiments of a decade earlier in part because of overall improvements in the quality of filmmaking and projection, but mostly because in the late 1920s, movies had stopped being silent. With the introduction of synchronized sound, filmmakers were able to trick two senses at once and overwhelm audiences with evidence that the unreal thing they were watching and hearing was actually happening. "Sound is a great factor in heightening the illusion," the *Times* noted. "The seltzer-squirting episode, for example, is doubly effective, because you hear the zizzz and the splash when it strikes." The film was a commercial and critical success and was nominated for an Academy Award for Best Short Subject (Novelty)— though it lost to a documentary about an explorer who flew a biplane over the peak of Mount Everest. MGM went on to commission two sequels, *The New Audioscopiks* (another collection of short demo clips) and *Third Dimensional Murder* (an original comedic narrative about a man who visits a spooky castle and is attacked by various scary creatures, including a witch, a skeleton, and Frankenstein's monster).

While 3-D movies reached into theaters, other forms of stereoscopic media worked their way into workplaces and homes. In 1938, German inventor William Gruber developed a new method for photographing 3-D images using two consumer-grade cameras loaded with Eastman Kodak Company's newly released Kodachrome color film. Gruber would shoot stereoscopic pairs of an image, print them onto fingernail-sized pieces of the translucent film, and then mount the two photos on opposing sides of a fist-sized paper disk. He also created a handheld stereoscope specially designed to view the little film reels—a device resembling a small pair of binoculars, with a small lever that would advance the reels from one image to the next.

When Gruber went on a trip to photograph the Oregon Caves National Monument with the rig, he was stopped by Harold Graves, the owner of a company that printed souvenir photographic postcards, who wanted to know why he had two cameras on one tripod. Gruber explained the invention, and shortly after, the two men went into business together to mass-produce and market the device. It made its public debut at the 1939 New York World's Fair as the View-Master, and within decades it would become the most commercially successful stereoscopic device in history.

Gruber and Graves originally conceived of the View-Master as a high-tech alternative to postcards, an immersive way to experience faraway places, and sold it alongside film reels depicting locations like national parks. But when World War II began, another big market opened up. Between 1942 and 1945, the US Department of War purchased over 100,000 viewers and millions of slide reels to help teach soldiers to recognize ships and airplanes. Each reel included stereo photos of friendly and enemy craft at different ranges and altitudes, and in front of different backgrounds like blue sky or clouds.

After the war, all those vets returned home and bought the now familiar gadget for home entertainment or as a gift for their kids. And when Graves's company obtained the rights in 1951 to make reels featuring Walt Disney's characters and amusement parks, the View-Master was solidified as a cultural icon. Current parent company Mattel says that since the product was invented, consumers have purchased over 1.5 billion stereoscopic reels.

View-Masters weren't the only entertainment gadget that caught on after World War II. Television was invented in the late 1920s, but the first consumer products were huge, expensive devices with tiny screens. In the late 1940s, the technology was finally ready for prime time—and with millions of soldiers settling down, buying homes, and starting families, TV sales boomed. By 1950, about one out of every ten American households owned a television. A year later, that number was one out of every four.

The television boom was good for electronics manufacturers and broadcasting companies, but bad for Hollywood. The days where everyone flocked to the "picture shows" were over; in order to attract customers back to the theaters, movie studios needed to tempt them with experiences they couldn't get in their living rooms. So they launched a whole new generation of cinematic 3-D experiences.

American filmmaker Arch Oboler kicked off the "golden era" of 3-D movies with the 1952 film *Bwana Devil*, an action-adventure movie based on a true story of man-eating lions that preyed on workers building a railroad in Kenya. Unlike the anaglyphic 3-D movies of the 1920s, *Bwana Devil* was presented via a polarized 3-D system: instead of lenses that filtered the wavelength of light into different colors (like red and green), the Natural Vision system used lenses that filtered the orientation of light waves into different polarizations (like whether they oscillate up and down or side to side). When the audience wore spectacles with two different polarized lenses—essentially, a pair of sunglasses—they could view a film in three dimensions and in true color.

Posters for *Bwana Devil* crowed that the new color 3-D was a "miracle of the age!" and promised "a lion in your lap!" and "a lover in your arms!" But initial reviews focused less on the eye candy and more on the eyeglasses. An article in the December 15, 1952, edition of *Life* magazine featured a now iconic J.R. Eyerman photograph of the film's premiere at Hollywood's Paramount Theatre—an auditorium full of elegant men in suits and women in dresses wearing goofy-looking cardboard sunglasses. "These megaloptic creatures," the caption stated, "looked more startling than anything on the screen."

Audiences loved the movie anyway. *Bwana Devil* went on to make more than $2.5 million at the box office, more than five times its production cost, and enough to count it among the ten biggest films of 1953. Soon all of Hollywood was scrambling to get more 3-D films into theaters.

Much of what they produced were low-quality B movies. 1953's *Robot Monster*—a film about an evil creature from the moon, portrayed by an actor wearing a gorilla suit with a fishbowl "space helmet"—is widely regarded as one of the worst movies ever made. But there was so much demand for 3-D that the film made money anyway, earning over $1 million in box office on a tiny $50,000 budget. And the 3-D fad did produce some critical successes, especially in genres like sci-fi (*It Came from Outer Space*), monster movies (*Creature from the Black Lagoon*), and horror. André De Toth's 1953 film *House of Wax* earned $23.8 million globally and influenced an entire generation of horror filmmakers; in 2014, the US Library of Congress added it to the National Film Registry, an elite selection of films recognized for their cultural, historic, or aesthetic significance.

House of Wax is also notable because it was the first 3-D feature to use stereophonic sound—all the audio in the film was recorded on multiple tracks and played on speakers set up on the sides, front, and rear of the auditorium, so that sounds seemed to come from the direction of an action. Paired with a 3-D picture, the stereo audio created an unprecedented sense of immersion; when the titular wax museum went up in flames, the sight and sound of crackling fire surrounded the viewer. "The result has a kind of spellbinding effect on the audience, giving a feeling of realism to a completely unreal story as well as a sense of participation," according to an article in a 1953 issue of *The Hollywood Reporter.*

The 3-D film fad burned hot and burned out fast. Hollywood movie studios produced more than sixty 3-D movies in the first half of the 1950s, but when they were shown in public, the experience was often marred by technical issues caused by sloppy projection. By 1954, the golden age of 3-D movies was already ending, and John Norling, one of the producers of *Audioscopiks,* wrote an epitaph for the era in the magazine *International Projectionist.* "What does the future hold for 3-D? Nothing but interment unless the industry realizes its great potential and supports the research and development that will assure the perfection and convenience required," he wrote. "The full possibilities of 3-D have not been explored."

Chapter 2

THE ULTIMATE DISPLAY

Hollywood didn't give up on immersive cinema when the 3-D fad was over. It just focused on another aspect of the theater experience that consumers couldn't get watching TV in their living rooms—very big screens. One of the biggest was developed by Fred Waller, an engineer and former head of Paramount Pictures' special effects department, after he noticed that shooting scenes with a wide-angle lens seemed to create a kind of three-dimensional effect in a film. It made him realize that filmmakers had been concentrating on duplicating stereopsis, but that other visual cues also contribute to depth perception, like peripheral vision. So Waller figured he could increase immersion if he made a screen that filled the observer's entire field of view.

He spent fifteen years working on the problem before creating his most famous invention: Cinerama, a theatrical experience with a screen so big it could present "a photographic view of the scene as a human pair of eyes would see it." The system used three projectors shining on different thirds of a convex screen that was

41

300 percent wider and 50 percent taller than a standard theater screen, stitching together a massive picture almost as big as the entire range of binocular vision. A Cinerama production would wrap around the audience and make them feel like they were literally inside the movie, not just staring at a picture framed on a wall.

In September 1952, Waller's company debuted a full-length film called *This Is Cinerama* at the Broadway Theatre in New York City. The 115-minute movie was a vivid showcase for the new technology, featuring scenes from around the world, including a canal tour of Venice, a bullfight in Madrid, and the finale of *Aida* performed at La Scala in Milan. An article in the magazine *Popular Mechanics* described one segment of the film, a boat ride through the Cypress Gardens amusement park in Florida, as so real the audience reacted physically, leaning sideways as the boat tipped and ducking so they wouldn't smack their heads as it went under a footbridge.

This Is Cinerama was an unqualified hit in New York, selling out for eight months at the Broadway Theatre before transferring to the newly rechristened Warner Cinerama Theatre in Times Square, where it ran for another year and a half. A handful of theaters in other major cities also converted to the format, and the film went on a traveling "road show" to sold-out crowds. Ultimately, the cost and complexity of the system kept it from entering widespread use—fewer than a dozen films were made using Waller's three-projector setup. But some of the people who saw Cinerama found that the experience was hard to forget.

———

One of those people was Morton Heilig, a Hollywood cinematographer who was inspired to do better. The problem with systems like Cinerama, he decided, was that they weren't ambitious enough. Humans have five senses, but movie theaters engage only two of them. And besides, he figured, they don't even engage those senses fully. What good is a curved screen that fills your field of view if you can see the edge of it when you turn your head, and what good is stereophonic sound if you can still hear a distracting conversation from the couple sitting in the row behind you in the theater?

"Every capable artist has been able to draw men into the realm of a new experience by making (either consciously or subconsciously) a profound study of the way their attention shifts," Heilig wrote in a 1955 essay called "The Cinema of the Future." "Like a magician he learns to lead man's attention with a line, a color, a gesture, or a sound. Many are the devices to control the spectator's attention at the opera, ballet, and theater . . . but the inability to eliminate the unessential is what loosens their electrifying grip."

What was needed, Heilig argued, was a new kind of entertainment, one that could surround viewers in a simulation that blocked out distractions and engaged all their senses, allowing them to experience the world "in all its magnificent colors, depth, sounds, odors, and textures." So he resolved to build an immersive media device that could do just that.

First, Heilig designed a stereoscopic headset—basically, a modern version of the Holmes scope, but with built-in television tubes that displayed a moving 3-D picture, instead of photographs printed on a piece of cardboard that showed a still image. Patented in 1960, the Telesphere Mask is now regarded as the first modern head-mounted display, a direct ancestor of twenty-first-century virtual reality hardware. But the device never worked perfectly, so Heilig only built a prototype and never took the invention into production.

Next, Heilig decided to build something bigger—a kind of immersive one-person theater he called Sensorama. Patented in 1962, the device consisted of a single seat facing a wraparound hood; users sat down, grabbed on to handlebars, and pressed up against a viewer that looked something like the inside of a high-tech hockey mask. The eyes were binocular lenses, allowing the user to watch a wide-angle 3-D movie. The mouth was a small grille that the device could blow air through in order to simulate wind. The nose was another vent, this one equipped with a variety of scents that could be released to match the setting of a film. Small speakers on either side of the hood played stereo sound effects to further the illusion, and the whole rig could even vibrate and tilt to simulate motion.

Heilig produced several short films to demonstrate the simulator's capabilities, including rides on a helicopter, a dune buggy, and a motorbike cruising through New York City. Writer Howard Rheingold described the motorbike ride after he tried one of the last working Sensorama devices in the mid-1980s. "I sat down, put my eyes and ears in the right places, and peered through the eyes of a motorcycle passenger at the streets of a city as they appeared . . . [from] the driver's seat of a motorcycle in Brooklyn in the 1950s. I heard the engine start. I felt a growing vibration through the handlebar, and the 3D photo that filled much of my field of view came alive, animating into a yellowed, scratchy, but still effective 3D motion picture," he wrote in his 1991 book *Virtual Reality*. "Sensorama was a bit like looking up the Wright brothers and taking their original prototype for a spin."

After the motorcycle ride, Rheingold found himself riding in a convertible with a young blond woman, listening to a pop song playing on the car radio. In the simulation's finale, he watched a belly dancer perform "with a come-hither look," a scene that originally featured the smell of cheap perfume released out of the nose vent whenever the woman came close to the viewer.

Though Heilig's first demos seemed to position Sensorama as a goofy diversion, they were intended to catch the attention of investors who might fund development and marketing of the device for more serious applications. In Sensorama's patent filing, Heilig described the invention as a tool that armies might use to train soldiers without subjecting them to the hazards of warfare, or that businesses could rely on to teach employees how to use heavy machinery. He even suggested that Sensorama could help solve the growing problem of overcrowding in American schoolrooms: "As a result of this situation, there has developed an increased demand for teaching devices which will relieve, if not supplant, the teacher's burden . . . accordingly, it is an object of the present invention to provide an apparatus to simulate a desired experience by developing sensations in a plurality of the senses."

Heilig's invention was ahead of its time, and he struggled to find investors or customers who understood its potential. In a bid to find funding to continue his research, Heilig pitched Sensorama to vehicle manufacturers including the Ford Motor Company and International Harvester, positioning the device as an interactive showroom display that could entice customers with the simulated experience of driving a new sedan or tractor. Neither company was interested. Eventually, Heilig put his machines into amusement arcades located in tourist destinations like Times Square and Universal Studios. At first the scheme worked, and the machines started earning income—but due to their complicated design the machines broke easily and spent more time out of order than they did collecting quarters.

Sensorama might have been doomed by its delicate construction and the high cost of producing 3-D movies, but Heilig didn't give up on his vision. In 1969, he patented a room-sized version of the device, called the Experience Theater, meant to provide the same multisensory experience to an audience of dozens or hundreds of people. The patent describes a huge concave screen, chairs that move and rumble, a system of blowers for simulating wind and distributing aromas, and special polarized glasses that allowed the audience to view a movie in three dimensions.

Though the Experience Theater also failed to reach production, it might sound familiar to anyone who's ever visited an amusement park and gone on a motion simulator. Modern Disney park rides like Star Tours or Soarin' look an awful lot like Heilig's invention, even though his work was completed decades before those attractions opened. In fact, while very little of Heilig's work ever came to fruition, time has shown the prescience of his vision—his work led the way toward modern virtual reality and inspired many other inventors to chase their own dreams of immersive, interactive environments.

———

While Morton Heilig was trying to find ways to make filmed pictures look more like the real world, other researchers were trying to teach computers to create convincing artificial environments.

In 1961, electrical engineer Ivan Sutherland began developing Sketchpad, the first interactive computer graphics program, as part of his doctoral thesis at the Massachusetts Institute of Technology. The system allowed users to draw with a light pen on a cathode ray tube, and the digital images they created could be stored, duplicated, or manipulated on the screen. Sketchpad made it possible to visualize and create highly precise diagrams, so Sutherland essentially invented the modern industrial art of computer-assisted design.

That alone would have been enough to consider him one of the forefathers of virtual reality. But then, in 1965, while he was working as a professor at Harvard University, Sutherland published a paper titled "The Ultimate Display," which predicted a whole series of innovations that would become critical to the future of VR, including head-mounted displays, eye tracking, motion sensors, and gesture-based controls. Through the use of these technologies, Sutherland imagined creating simulations so convincing that they were indistinguishable from real life—or better yet, simulations of things that don't actually exist in real life.

"A display connected to a digital computer gives us a chance to gain familiarity with concepts not realizable in the physical world," he wrote. "For instance, imagine a triangle so built that whichever corner of it you look at becomes rounded. What would such a triangle look like? . . . There is no reason why the objects displayed by a computer have to follow the ordinary rules of physical reality with which we are familiar. The kinesthetic display might be used to simulate the motions of a negative mass. The user of one of today's visual displays can easily make solid objects transparent . . . Concepts which never before had any visual representation can be shown."

Decades before the term *virtual reality* would actually come into usage, Sutherland envisioned exactly that: immersive computer-generated displays, artificial environments with their own rules, and virtual spaces where users could see and do impossible things. Sutherland even imagined an extension of this technology where the virtual became physical—an idea that would enter popular culture twenty-two years later as the holodeck on *Star Trek: The Next Generation.*

"The ultimate display would, of course, be a room within which the computer can control the existence of matter," Sutherland wrote. "A chair displayed in such a room would be good enough to sit in. Handcuffs displayed in such a room would be confining, and a bullet displayed in such a room would be fatal. With appropriate programming such a display could literally be the Wonderland into which Alice walked."

In 1966, Sutherland started building hardware that would let users take a trip down the rabbit hole. His head-mounted display consisted of an unwieldy helmet with miniature CRT monitors on either side, and binocular lenses pointed at semi-transparent mirrors that reflected each screen. When a user strapped on the device and peered through the goggles, they could see simple wire-frame graphics overlaid on the real world. Two sensors, one mechanical and the other ultrasonic, measured the position and the direction of their gaze, allowing the computer to update its graphics as the user moved and looked around the room. And because all this gear made the headset unbearably heavy, it was suspended on wires from a rig attached to the ceiling—a setup that earned it the nickname the "Sword of Damocles," after the legendary weapon that dangled precariously over a Sicilian king's throne.

In his first experiments with the hardware, Sutherland produced a simple wire-frame cube that appeared to float in midair in front of the viewer, who could walk around it and examine it from any side. Later experiments put the user inside the cube, drawing a square "room" around them with walls, windows, and a door. "Even with this relatively crude system, the three-dimensional illusion was real," Sutherland reported.

———

Meanwhile, academics weren't the only people interested in the potential application of virtual reality. The US military was quick to realize that the technology could be put to use on a battlefield, and in 1967 a twenty-three-year-old second lieutenant at Wright-Patterson Air Force Base in Greene County, Ohio, started a project that would eventually earn him the nickname of "the grandfather of virtual reality."

Thomas A. Furness III was born in 1943 and raised in Enka, North Carolina, a tiny factory town that was about as different as possible from the high-tech worlds he'd later help invent. But young Tom took an early interest in science and engineering, and by the time he was in grade school he'd taught himself how to build and repair all kinds of electronic gadgets. His teachers would let him tinker in the classroom while other students covered material Furness had learned long ago; once a week, they'd even hand over the class for "Tommy Time," and the precocious boy would spend half an hour instructing his own schoolmates and showing off his latest projects. When he was fourteen, the Soviets launched Sputnik into orbit, and Furness decided he wanted to build rockets and travel into space. His junior-year science fair project, a working rocket telemetry system, won an award sponsored by the US Navy at the North Carolina state science fair. When he graduated from high school, he enrolled at Duke University, joined the Air Force Reserve Officer Training Corps, and earned a degree in electrical engineering.

After college, the Air Force sent Furness to Dayton, where engineers were working on ways to improve the high-tech combat aircraft the United States desperately needed as it escalated the war in Vietnam, and assigned him to the Armstrong Aerospace Medical Research Laboratory, or AAMRL.

"My job was to figure out, how in the world do we interface humans with machines?" Furness said. "We had a problem of complexity . . . we had aircraft that had fifty computers on board, and just one operator. How in the world was he going to make sense of all that?"

Furness realized that following the invention of the digital computer, American engineers had "improved" the nation's fighter jets with new systems to the point that their operators were completely overwhelmed, sealed in cockpits so full of hardware they could barely see out the windows, and faced with systems so complex they could never process all the information at hand. When one pilot gave him a facetious drawing of the "pilot of the future"—a man with six arms so he could handle all the controls—Furness resolved to completely rethink the interface between aircraft and operator.

At first, Furness focused on creating helmet-mounted displays, like Ivan Sutherland's "Sword of Damocles," that could superimpose computer graphics onto a pilot's view of the real world. By tracking head movements, the system could also sense where the pilot was looking, and dynamically update to show information relevant to the system in question or the task at hand. Early prototypes showed promise, so in the 1970s, Furness decided to take the idea even further and develop an entirely virtual cockpit, where pilots could be completely immersed in a computer-generated version of the world.

Furness's project, known as the Visually Coupled Airborne Systems Simulator, or VCASS, was completed in September of 1982. The centerpiece of the system, a headset nicknamed the "Darth Vader helmet," consisted of two CRT displays—one for each eye—hanging from a platform and connected to eight mainframe computers running computer graphics software. VCASS filled several rooms and used so much electricity that Furness joked he "had to tell Dayton Power and Light" whenever he was going to power up the system.

Once it was running, VCASS's stereoscopic displays would show a simple, cartoonish landscape, like an early 3-D video game, that surrounded the user and would update instantly when the user looked around or turned his head. Real-time information on the aircraft's systems appeared in the virtual environment as needed, and simple icons on the landscape represented everything from airports to enemy fighter jets. The system included eye-tracking hardware, motion sensors, and even a voice input system: pilots could fire a missile by simply speaking a command to the computer.

The Darth Vader helmet offered a radical new way for fighter pilots to control their aircraft. But what Furness created was much more than just a new avionics system: it was a profound new interface between man and machine. Inside VCASS, humans could interact with computers using the same senses, skills, and instincts they used in their everyday life. It was simple, immersive, and easy to understand—so easy that Furness said totally untrained users, including his high school–aged daughters, would strap into the system and learn how to fly a fighter jet in no time at all.

"It was amazing," Furness said. "We found that instead of looking at a screen . . . it was like we were pulled into another world. The transformation was remarkable. When we did that, all the abilities that we used in the real world could actually now be used in this computer-generated world, which meant that users had much more power, much more bandwidth to and from the brain."

———

Pilots flying inside of one of Furness's rigs might seem to exist inside a virtual reality, but their interactions with the aircraft were still very physical—they had to steer with a control yoke, push pedals to move the rudder, and press all kinds of buttons and switches to perform tasks ranging from turning on the radio to firing a surface-to-air missile. What virtual reality still needed was a way to interact with completely virtual objects—to see something in the simulation that didn't actually exist, and then be able to grab, manipulate, or move it.

Early efforts focused on developing data gloves, high-tech handwear that could track the exact movements of its wearer's fingers. The first such device was developed in 1977 by University of Illinois scientists Daniel J. Sandin and Thomas Defanti. It was named the Sayre Glove after a colleague, Richard Sayre, who had given them the idea for the project. The glove was made with light-conducting tubes running along the top of each finger, a light source on one end, and a photocell on the other. When the user bent a finger, it reduced the amount of light that made it to the sensor, allowing a computer to approximate the amount that the finger was flexing. It was lightweight, easy to produce, and inexpensive, but not terribly accurate—really only useful for flipping virtual switches or moving virtual sliders.

In 1982, Thomas Zimmerman, a scientist working at the Atari Research Center in Sunnyvale, California, picked up the research and started building a data glove for his own pet project. At the time, Atari was one of the world's biggest electronic entertainment and video game companies, and Zimmerman had an idea for a new gadget that would allow people to play "air guitar" and really make music—a glove that could track its user's hand and finger positions as they pretended to work the frets and strum the strings of an imaginary instrument. Connect it to an electronic synthesizer, and pretending to play would produce real live music.

While he was working on the project, Zimmerman became friends with another coworker at the Atari Research Center, an engineer named Jaron Lanier who was also a fan of music. In 1983, they teamed up to improve the design of the data glove, and over the next two years made so much progress that they would go on to found a new company that gave birth to the modern virtual reality industry: VPL Research.

Chapter 3

CONSOLE COWBOYS

Jaron Zepel Lanier, the man who coined the phrase *virtual reality,* was born in 1960 in New York City, but when he was just a baby his parents fled bohemian Greenwich Village and moved off the grid to El Paso, Texas, along the US border with Mexico. His mother and father, Jewish survivors of a concentration camp in Austria and a pogrom in Ukraine, didn't trust the government and wanted to "live as obscurely as possible." Lanier's mother died in a car crash when he was only nine, and not long after, a series of illnesses forced him to spend nearly a year in a hospital. When he emerged and returned to school, he was overweight, socially isolated, and a favorite target of local bullies. The few friends he made were outsiders and "oddballs"—including a radar technician from nearby Fort Bliss who he met in the aisles of a RadioShack while browsing through drawers of transistors and capacitors. The young soldier took the boy under his wing and taught him basic electronics.

After Lanier's home was destroyed in a fire, his father moved them to the remote village of Mesilla, in the high desert of New Mexico. Young Jaron and his dad lived in a tent and spent their free time building a huge house consisting of multiple interconnected geodesic domes. "The overall form reminded me a little of the Starship *Enterprise*," Lanier wrote. It took seven years to complete, but in the process, Lanier developed a taste for creating weird and wonderful environments.

Over time, Lanier made more unusual friends, including his neighbor Clyde Tombaugh, a sixtysomething astronomer who discovered the dwarf planet Pluto in 1930 and worked as the head of optic research at the White Sands Missile Range. Tombaugh taught him how to grind lenses and mirrors, helped him build a telescope, and introduced him to the high-tech computers at the nearby army lab.

Precocious, brilliant, and restless, Lanier dropped out of school at age fourteen and started taking classes at New Mexico State. (He hadn't graduated from high school or been admitted to the university, but because of his intelligence and sheer bravado, no one ever bothered to turn him away.) He took music classes, learned about composition and orchestration, and started to think about using computers to make music of his own. He also studied programming and computer graphics and, when he discovered Ivan Sutherland's research into the Ultimate Display and the Sword of Damocles headset, fell in love with the idea of creating virtual worlds.

"Reading about Ivan's work was challenging for me, because each sentence took me by storm," Lanier wrote in his book *Dawn of the New Everything: Encounters with Reality and Virtual Reality*.

"You would eventually be able to make any place and be in it via this device . . . plus, other people could be in there with you. . . .

"I was fifteen years old and vibrating with excitement. I had to tell someone, anyone. I would find myself running out the library door so that I didn't have to keep quiet; rushing up to strangers on the sidewalk . . . 'You have to look at this! We'll be able to put each other in dreams using computers! Anything you can imagine! It's not just going to be in our heads anymore!'"

A few years later—following stints as a goat farmer, art student, and midwife—Lanier moved to Silicon Valley, found work composing music and sound effects for the nascent video game industry, and eventually programmed games of his own. In 1983, he made an interactive music-making game called Moondust for the Commodore 64 home computer, and the sales produced enough profit for him to set up a garage workshop and get to work on another pet project.

Lanier's new idea was a computer programming language for the masses, a piece of software called Mandala that used graphics and sound instead of esoteric lines of code. This visual programming language took the rough form of a piece of music. Users arranged icons on a screen like notes on a musical staff, and their relative positions and order would tell the computer what to do. It was an exciting idea, and in September 1984, when *Scientific American* published a special issue about computer software, the editors decided to use an illustration of the Mandala language on the magazine's cover.

Shortly before the issue was due to go to print, an editor called Lanier on the phone and told him they needed to know which university or company he was affiliated with, in order to give him proper credit in the magazine. Since Lanier was working solo, he decided to make something up.

"I said, 'Oh, VPL, standing for Visual Programming Languages, or maybe Virtual Programming Languages,'" Lanier told a reporter in a 1993 *Wired* magazine interview. "Mainly it was just this spontaneous thing to get this guy off the phone. And then I told him to put a comma and 'Inc.' after it and never gave it a second thought. And then when the issue came out months later, all these people called up wanting to invest!"

Flush with interest in his project, Lanier doubled down on development of Mandala, but he had a big problem: Since the new programming language depended heavily on graphics and imagery, it required the use of large, colorful, sophisticated displays. A standard computer terminal of the era wouldn't be adequate for the task. Furthermore, since the exact placement of each icon on the screen directly translated into the language's instructions, Mandala required a simple and accurate way to move graphics around. Standard keyboards weren't right for the job, and even though the computer mouse had been invented decades earlier, in the early 1980s the pointing device wasn't in wide use.

Lanier's solution to these problems was to return to Ivan Sutherland's idea of a virtual display. A head-mounted 3-D system like the Sword of Damocles could present a huge, immersive canvas for users to paint Mandala's pictorial arrays. And to complete the interface, some kind of hand-tracking implement would work best—perhaps a glove like the one his friend Thomas Zimmerman was building to allow people to play computerized air guitar.

So Jaron Lanier's imaginary company became reality. With the help of Zimmerman and a few other friends, VPL Research began developing new hardware—a head-mounted display and a wired glove for viewing and manipulating objects in a virtual space.

"All of a sudden we had a company," Lanier said. "It was just something I fell into, it was crazy." Initially, he thought that hardware would be a side business, and VPL's primary focus would be to develop his programming language. But as soon as people saw the simulator they were building, it was all they cared about.

"Potential investors would come around and I would show them this thing, and I'd say, 'Now look at this neat language,' and they'd say: 'Language! You're using a glove! My God!' So suddenly the whole focus shifted," Lanier said.

At some point, Lanier coined the term *virtual reality* as a marketing term to describe what VPL was building—or perhaps, he admits, he heard the phrase somewhere else and then repeated it until it entered popular use. The company also branded its headset (the EyePhone) and its controller (the DataGlove), and started work on a full-body input device (the DataSuit). VPL was creating a new industry from the ground up.

In October 1987, just a few months after VPL released a commercial version of the DataGlove, a photograph of a hand wearing the high-tech gadget appeared on the cover of *Scientific American*. A line of text explained that the subject of the issue was "the next revolution in computers"—and that advanced interfaces like the DataGlove would "transform computing into a universal intellectual utility."

The hype was on.

———

The world was primed and ready for someone to invent a way to enter the world of computers. In 1982, Walt Disney Productions released *Tron*, a science fiction adventure film about a software engineer who gets sucked into a mainframe and must do battle with evil programs. The film was well received by critics, and its depiction of a glowing high-tech computerized landscape helped it develop an immediate cult following. Then in 1984, science fiction author William Gibson further popularized the idea of virtual reality when he published his dystopian science fiction novel *Neuromancer.* In that story, computer hackers known as "console cowboys" transfer their disembodied consciousness directly into a "consensual hallucination" known as the Matrix in order to steal data and sell it for profit.

When actual products like the DataGlove and EyePhone started hitting the market, these fictional depictions of futuristic realities suddenly seemed excitingly plausible, and the next big revolution in technology following the invention of the personal computer: a May 1987 article in *The Washington Post* declared that virtual reality researchers were "doing nothing less than inventing a new relationship between humans and machines."

More businesses rushed into the space to fill demand for VR products, including Autodesk, a $100 million software firm best known for making a computer-aided design application called AutoCAD. Inventor Eric Howlett founded a start-up called Pop-Optix Labs, and worked with the NASA Ames Research Center to develop a 1985 VR demonstration called the Virtual Interface Environment Workstation, or VIEW; in March 1989, that company started selling a version of his hardware as the LEEP Cyberface, one of the first commercially available head-mounted displays.

Even toy companies wanted in on the action. In 1987, Mattel released the Power Glove, a $75 wearable controller for the Nintendo Entertainment System video game console. The accessory was based on VPL Research's DataGlove and codeveloped by the company. It sold poorly, but its prominent appearance in the Nintendo-produced adventure comedy film *The Wizard* helped hook the interest of an entire generation of young gamers, getting them thinking about how cool it would be to manipulate objects in virtual reality.

Meanwhile, Tom Furness's lab at the Wright-Patterson Air Force Base had boomed to include nearly a hundred researchers, and the increased visibility of VR products in the marketplace and on the news meant that his once-obscure military projects suddenly were attracting interest from all kinds of industries.

"The most remarkable thing was the phone calls I started getting," Furness said. "A mother called me and said . . . 'My child has cerebral palsy. Is there anything you can do with that technology to help my child?' A surgeon called me . . . he said, 'Is there any way you can help me with navigating inside the body? I have all this information from a CT scan, but it's on a light box on the wall. Is there any way you can project that into the patient so I can find my way around?' . . . I was getting three or four phone calls a week from people thinking about the applications of this technology."

So Furness decided to beat his swords into plowshares, and in 1989 he left the Air Force and started the Human Interface Technology Lab at the University of Washington. "I realized we needed to get this technology out in the world where it can really do some good," he said. "I wanted to build upon all we had learned."

With so much new development under way in the nascent industry, researchers like Furness and Lanier were in high demand in the media, and VR became a hot subject of discussion on TV news shows and in newspaper and magazine articles. A February 1990 article in the *Chicago Tribune* proclaimed that Lanier "may someday send the world crashing through the looking glass." In April, an article in *The Guardian* crowed that "we are witnessing the birth of Virtual Reality—the total electronic environment which will change our perception of the real world as surely as books or television."

But excitement for a product doesn't necessarily translate into revenue. In 1990, if a company wanted to use VPL's products, even a single installation would be hugely expensive: $8,800 for the DataGlove and its tracking sensors, $9,400 for the EyePhone headset, and $7,200 for the necessary software. If you really wanted to go all in, you could buy VPL's flagship product: an all-in-one system called RB2, or "reality built for two," which included all the hardware, software, and accessories needed to set up a VR environment where two users could interact with each other. It cost $45,000. Truly advanced users might want to drop another $50,000 for a full-body DataSuit. And none of these prices included the massive cost of the high-end computer graphics workstations required to run a virtual reality simulation—anywhere from $75,000 to a quarter of a million dollars or more, depending on how powerful the system.

Unsurprisingly, there weren't that many customers. VPL sold a few thousand DataGloves and EyePhones to fellow VR developers, researchers working on cutting-edge applications like remote surgery, and governmental clients like NASA and the US Department of Defense. But demand was never high enough to cover the immense costs of development and production, which included a brand-new factory in Silicon Valley to produce the company's uniquely high-tech products.

It didn't help that CEO Lanier had little experience in business and was actually the youngest person at the company. He described himself as a "babe in the woods," and made frequent beginner's mistakes that cost the company dearly: "One time I signed an important contract for VPL that I thought had been vetted by the lawyers without checking the backside of the paper, where the other party had snuck in additional language that ended up screwing us," he wrote in his memoir.

By the end of 1990, even though virtual reality was a hot emerging technology, VPL ran out of money, and while the company kept operating, it had to file for bankruptcy protection.

———

But even if VR businesses were struggling, the ideas be-hind virtual reality continued to gain ground in popular culture. In 1991, Howard Rheingold published a book about the industry, *Virtual Reality: The Revolutionary Technology of Computer-Generated Artificial Worlds—and How It Promises to Transform Society*, and it became a bestseller thanks to readers clamoring to find out more about the emerging technology.

In 1992, VR went fully mainstream as the subject of a high-profile science fiction horror film called *The Lawnmower Man*. The movie—which was advertised as based on a Stephen King

short story but had very little to do with that source material—
follows future James Bond actor Pierce Brosnan as Dr. Lawrence
Angelo, a researcher who is programming chimpanzees to become super-soldiers using virtual reality training simulations.
When a test subject breaks out of his lab and kills several bystanders, Dr. Angelo shifts his focus to experimenting on his intellec-
tually disabled gardener, only to find that the VR treatments work
a little too well: The titular "lawnmower man" develops genius
intelligence, telepathic abilities, and eventually his own murder-
ous intentions. By the end of the film, the gardener has evolved
into "pure energy" inside virtual reality, and announces his presence to the world by causing every telephone on the planet to ring
simultaneously.

The movie was a modest hit, grossing over $32 million domes-
tically, even though reviews were terrible. Audiences were taken
by the special effects and vivid computer-generated depictions of
life (and sex) inside VR, and the film developed a cult following of
sci-fi fans and computer enthusiasts who were eager to try out the
technology.

Later the same year, a much more successful work of fiction
outlined another vision of life inside a virtual environment. In
Neal Stephenson's science fiction novel *Snow Crash*, users controlled personalized avatars inside a shared space called the
Metaverse. Unlike the depiction of VR worlds in *Neuromancer*
and *The Lawnmower Man*, which were largely abstract geo-
metrical spaces, the Metaverse was built around a hundred-
meter-wide road known as The Street that ran the circumference
of an otherwise featureless planet. The Street was lined with

buildings, houses, and businesses, and users experienced it in first person just as they would an actual urban environment. Their avatars shopped, visited bars, and spent time with friends. There was more to this world than just hacking computer networks and fighting security programs—the Metaverse imagined a vivid alternate reality, a place where ordinary people might want to visit.

Consumer interest in the technology continued to increase, and peaked in 1995 when nearly a dozen VR-centric films (including *Johnny Mnemonic, Virtuosity, Strange Days, Cyber Bandits, Virtual Combat,* and *Terminal Justice*) were released into theaters, and a TV series called *VR.5* debuted on the Fox network.

The movies sold tickets, but the products continued to fall flat. Occasionally, excessive costs killed them in their infancy. In the early 1990s, Hasbro spent $59 million and more than three years developing a console and headset called the Home Virtual Reality System, before it gave up. At the time, CFO John O'Neill told the Associated Press that the gadget's $300 price tag would have priced it out of the consumer market.

More often, VR was doomed by technical problems. In 1996, Nintendo released a $180 video game console called the Virtual Boy, but its promise of 3-D graphics fell flat. The headset's red monochrome display, low resolution, and use of high-speed vibrating mirrors gave its users neck pains, dizziness, nausea, and headaches. Nintendo sold fewer than 800,000 units and discontinued the product after only a year.

In 1998, continually plagued by high costs, low sales, and unreliable manufacturing, VPL Research filed again for bankruptcy.

This time the company didn't recover; during the proceedings, all its patents were purchased by the Silicon Valley computer giant Sun Microsystems, which had its own plans to develop virtual reality hardware—but which soon gave up on the idea and never actually released any related products.

Even though the virtual reality industry died, the dream survived. In the final years of the twentieth century and into the first decade of the twenty-first, small communities of VR enthusiasts and collectors stayed in touch with one another on the Internet to share their research, boast about their hardware collections, and plan for the future. A few late-breaking cyberpunk movies— particularly the 1999 blockbuster *The Matrix*—introduced new audiences of young technologists to the excitement of virtual reality. And a new generation of massively multiplayer online video games gave players a taste of what cyberspace might be like in the future—always connected, with users everywhere, and presented in high-fidelity 3-D graphics. All they needed was a way to get the games off the screen and into the world around them.

Chapter 4

INTO THE RIFT

It's dark and creepy in the cargo hold of the *Sevastopol*; something could be hiding in here, and I'd never know it. It makes me nervous, and as I walk between the piles of crates, I stop and peek around every corner.

A noise makes my heart skip a beat. I tell myself it was just dripping water.

I shouldn't be scared, since I know this is just a video game— a demo of a first-person horror adventure set in the world of Ridley Scott's *Alien* film franchise. But I'm wearing a virtual reality headset called the Oculus Rift, and the Rift makes it real: The game fills my entire field of vision, and when I turn my head to look around, the world moves with me. It feels like I'm actually on a space station being stalked by one of H. R. Giger's xenomorphs. And that's not a good feeling.

There's another sound, this time off to the left, and I turn to see a heavy blast door slide open at the end of a corridor. Behind it, there's a crouching, bipedal form, the size of a large man, covered in a shiny black exoskeleton.

The smart thing to do is run away. But inside the Rift's virtual world, I'm hypnotized by the spectacle—or as less kind observers might call it, paralyzed by fear. So I stare, frozen, as the alien rises, closes the gap between us, and wraps its arms around my body. Its dripping mouth opens, and a set of inner jaws plunge toward my face.

I twitch in horror and emit an involuntary squeal of panic. From behind me—this time in the real world—I hear a laugh. The creator of the Oculus Rift has been watching as I play. "You got eaten? You didn't last very long," he says.

Palmer Luckey has been waiting for a game like this since he was a kid—which wasn't that long ago. Just twenty-one years old as of this meeting, he started making virtual reality headsets when he was sixteen, and founded his company, Oculus VR, only three years later. In less time than most of us take to finish college, he's gone from tinkering in his parents' garage to doing what generations of technologists before him tried and failed to accomplish: bringing virtual reality to the masses.

━━━━━

The path to this new world began in a setting common to so many modern success stories, it's almost become a cliché: a garage in suburban California. But Luckey wasn't a striving Stanford graduate or dot-com refugee. He was a nerdy little kid who became obsessed with science fiction stories.

Born in 1992 into a middle-class family, Palmer Freeman Luckey grew up in a house on the ocean boardwalk in the Long Beach neighborhood of Los Angeles. His father, Donald, worked as a salesman and assistant manager at a car dealership; his mother, Julie, was a management consultant and stay-at-home mom for Palmer and his three younger sisters.

The Luckeys homeschooled their children and encouraged them to pursue their passions. Donald was an amateur mechanic and taught Palmer basic machining and electronics. "My dad taught me to solder," he says. "He wasn't very good at it, but he taught me. Mostly I learned from watching YouTube videos."

Inside his garage workshop, Luckey started small, custom-building his own computers. But before long he moved on to bigger—and often more dangerous—projects. "I burned myself a ton," he says. "I've blown off my eyebrows a few times." Once, when he was experimenting with a high-voltage Tesla coil, he caused a short circuit, nearly electrocuting himself and scarring his wrist with an electrical burn. On another occasion, he accidentally discharged a laser into his eye and burned a tiny blind spot on his retina. He says he doesn't notice it, though. As he'd learn later when he studied VR, the human brain is excellent at correcting errors in our vision.

A favorite pastime involved "modding" video game consoles—taking vintage hardware and modifying it with newer components or to fit into portable, battery-powered cases. When he was sixteen, he cofounded a website called ModRetro Forums to discuss the hobby with like-minded hobbyists around the globe.

"I'm a huge fan of online communities, and I don't think that I'd be where I was today without them," Luckey says. "If you have one person in their garage and they're trying to figure out how to do something, they're gonna have a long path ahead of them. But on ModRetro, there were a lot of projects where you'd have twenty, thirty people all working together. That kind of collaborative problem solving is really powerful."

Life outside of the garage revolved around playing video games (Super Smash Bros., Chrono Trigger, and GoldenEye 007 were among his favorites) and consuming science fiction (particularly cyberpunk stories like *The Matrix* and *The Lawnmower Man*).

"Virtual reality is in so much science fiction, across a wide variety of stories, that even if you're not particularly interested in VR, if you're a sci-fi enthusiast you end up learning a lot about it," says Luckey. "That's what happened. I grew up my whole life thinking virtual reality was very cool, and I thought that it must exist in secret military labs somewhere."

Inspired, Luckey went hunting for evidence of the arcane technology. He scoured eBay sales for outdated and abandoned bits of VR hardware, and slowly amassed an impressive collection. In one score, he bought a $97,000 headset for only $87.

To fund his efforts, he started a business repairing broken iPhones. "First I was just fixing individual people's phones," Luckey says, "but then I realized that a much better way to do it was to go on eBay, buy lots of five or ten broken iPhones, and then fix up as many as I could and resell them." The scheme made him more than $30,000, and he poured nearly all of it back into his hobby.

But Luckey wasn't satisfied with collecting other people's failed VR experiments. He wanted a system that worked, so he started pulling pieces from old gadgets made by companies like Vuzix and eMagin, and hacked them together into something original. "I was modifying existing gear really heavily, using new lenses, trying to swap lenses from one system into another," he says. "I built some shitty stuff."

With time, his work improved. In 2009—when he was only seventeen—Luckey started building a VR headset he called the PR1, or Prototype One. "The entire optical system was all custom for that head-mounted display," he says. "That was the first time where I tried to build something new from the ground up."

On November 21, 2010, Luckey announced his project in an online forum for 3-D gaming enthusiasts called MTBS3D, or "Meant to Be Seen in 3-D." He explained that he'd built the device into the shell of a MRG2.2 head-mounted display, originally manufactured in 1996 by Winnipeg-based Liquid Image Corporation, but heavily modified it with improvements including a new LCD screen, new lenses, and a new sound system. "I would go on and on and on about all the revisions I went through in the build process," Luckey wrote, "but I do not want to bore you."

———

You might expect a homeschooled kid who spent his child-hood hacking cell phones and inventing computer hardware to be awkward and introverted in conversation. But Palmer Luckey is charming, voluble, and funny. He's laid-back and casual, almost always dressed like he's come straight from the shore, in cargo shorts and brightly colored beachwear. The soles of his feet are often black with dirt from treading around barefoot; if he wears shoes at all, they're likely to be sandals.

He studied public speaking and opera as a kid, and aspired to work as a singing gondolier for tourists on Long Beach's Naples Island, but he didn't make it through training. "I was terrible," he says. He worked another part-time job cleaning the boats at a sailing center and, in his free time, raced dinghies. He also played golf, at least until another kid hit him in the jaw with a club, causing a compound fracture. There's a deep scar on his chin from that, one he liked to flaunt as a teen because he thought it made him look tough—though sometimes, he jokes, "if you catch me from the right angle, it looks like I have a crazy butt chin."

He's well rounded too, able to discourse knowledgeably on almost any topic, not just games and electronics. It's a good thing, since a conversation with him is likely to veer in as many different directions as L.A. traffic, from modern music ("Why doesn't anyone write power ballads anymore?") to why the fashion industry deserves more respect in tech circles ("Many engineers don't see fashion as cool because it's so form over function, but the advances that make empires like Spanx are very much about function").

And when it was time to start college, Luckey didn't automatically pursue a degree in science or engineering. In 2009, when he was seventeen, Luckey enrolled at California State University, Long Beach, where he planned to major in journalism. "I wanted to be a tech journalist who understood how the technology worked," he says. "I thought there was a place for someone who understood engineering."

———

When he had spare time between classes, Luckey kept working on his HMD prototypes. In the summer of 2011, his hobby helped him land a part-time job working with virtual reality pioneer Mark Bolas at his lab in the Institute for Creative Technologies at the University of Southern California. Bolas and his students had spent years refining hardware and software for VR headsets, and all their innovations were released on an open-source license. Working at the lab, Luckey absorbed both their wisdom and their technology, and quickly applied them to his own project.

Meanwhile, Luckey continued to post about his progress in the MTBS3D forums. On September 16, 2011, he announced a second headset, dubbed PR2; just over a week later he shared a list of improvements for PR3, an upgraded wireless version. At each step, his fellow enthusiasts offered encouragement and suggestions. In one thread, a forum member even offered to buy one of the prototypes. Luckey replied with an offer to lend him one for free, and demurred on the idea of starting a business. "If a lot of people are interested, maybe I could see about making more to sell (Maybe a kit?)," Luckey responded. "My goal is not really to make money, obviously, or I would not be telling everyone exactly how it works!"

But the conversation may have sparked something in the enterprising teenager. In April 2012, nineteen-year-old Palmer Luckey completed the sixth prototype of his home-brewed VR rig, and made a post on the MTBS3D forums requesting help with a plan to underwrite future development via a campaign on the crowdfunding website Kickstarter:

Hey guys,

I am making great progress on my HMD kit! All of the hardest stuff (Optics, display panels, and interface hardware) is done, right now I am working on how it actually fits together, and figuring out the best way to make a head mount. It is going to be be out of laser cut sheets of plastic that slide together and fasten with nuts and bolts. The display module is going to be detachable from the optics module, so you will be able to modify, replace, or upgrade your lenses in the future!

The goal is to start a Kickstarter project on June 1st that will end on July 1st, shipping afterwards as soon as possible. I won't make a penny of profit off this project, the goal is to pay for the costs of parts, manufacturing, shipping, and credit card/Kickstarter fees with about $10 left over for a celebratory pizza and beer. ;)

I need help, though!

1) I need something that illustrates the difference between low field of view HMDs and high FOV HMDs, probably some kind of graphic illustrating the difference in apparent screen size. Would probably want to compare the rap 1200VR, the HMZ-T1, and the ST1080. Maybe throw in a few professional HMDs like the SX111 for good measure.

2) Logo/s. I am listing the organization as "Oculus", I plan on using that name on my VR projects from here on out. The HMD itself is tentatively titled "Rift", if you have better ideas, let me know. I based it on the idea that the HMD creates a rift between the real world and the virtual world, though I have to admit that it is pretty silly. :lol:

3) Ideas for what I should show off in the Kickstarter video.

4) Ideas for Kickstarter rewards. The obvious one would be a full HMD kit, but I want to have some lesser monetary options for people who just want to show support. Laser cut badges? Some kind of software? On the other end, it seems like it would be a good idea to have some more expensive options that net you stuff like a wireless battery/video pack, or a motion tracker.

5) Anything else I am forgetting!

The help is appreciated! Really excited about this, I think it could be the kind of thing that jumpstarts a bigger VR community, and hopefully shows that there is a big demand for wide FOV, truly immersive displays.

The forum members rallied to the cause of the Oculus Rift, with offers of free design work, marketing help, and technical assistance. One commenter even urged Luckey to charge more and earn a profit: "It's extraordinarily cool of you to keep this rock-bottom on the pricing tier but you absolutely deserve to make something from your efforts. This could be the start of something much bigger and you needn't limit yourself unnecessarily."

And at least one of the interested forum members was no ordinary hobbyist. John Carmack had made his first splash in the video game business in 1991, when he cofounded id Software; over the following decade, he cemented a legendary reputation working as lead programmer on seminal games including Doom, Quake, and Wolfenstein 3D. Earlier in the month, Carmack had posted on MTBS3D asking for help modifying a Sony head-mounted display, and Luckey had responded. "We had a public discussion on why it would be very difficult," Luckey says. "There were other people like, 'Oh, my God. It's John Carmack.' But I wanted to play it cool."

A week after Luckey announced his Oculus Rift project, Carmack sent him a private message via the forum. "I think he had gone through and seen some of the threads I had posted about my work," Luckey says. "He asked me if he could buy or borrow one of my prototypes so he could check it out. I told him he could just borrow it for free." Luckey packed up a headset and shipped it to Carmack's lab in Texas.

On May 17, 2012, Carmack posted a review of the Rift on the MTBS3D forums. "After dialing everything in, this is by far the most immersive HMD of the five I have," he reported. "If Palmer comes close to his price target, it will also be the cheapest. I will be including full support for this in the next new PC title we release."

An endorsement like that from John Carmack was more than enough to get the forum's VR enthusiasts lining up to back Luckey's Kickstarter. But Carmack wasn't finished. The Rift prototype was great hardware but lacked decent software; the device ran a hodgepodge of programs originally intended for 3-D monitors and commercial projectors. So Carmack modified code he'd originally written for the 2004 first-person shooter Doom 3, allowing the video game to run on the headset. And then he brought the Rift to the Electronic Entertainment Expo in Los Angeles, one of the video game industry's largest annual trade shows.

At the event, Carmack ran demos of Doom 3 on the headset for dozens of high-profile game designers, executives, and journalists. Each of them saw an action-packed, three-dimensional, truly immersive simulation—and walked away a believer. "I felt like I was literally standing in the game, with full 360-degree vision and interactivity," a CNN reporter wrote about the experience. "The potential and payoff . . . was truly astounding."

———

To the layperson, the Oculus Rift might not look too different from the VR headsets they'd seen in science fiction movies.

If anything, the prototype looked cheap and goofy—a block of plastic held together by silver duct tape, attached to a head strap borrowed from a pair of Oakley ski goggles. So what made it special?

Before the Rift, VR headsets were big and heavy, packed with custom screens and complicated optics. They had small fields of view, so wearing one felt like looking at the world through a porthole. The images they displayed had poor contrast and low resolution. And the picture was laggy—turn your head and it took a moment for the system to react, creating a disparity between motion and vision that inevitably led to nausea.

But when Palmer Luckey set to work on the Rift in his parents' garage, he did so with limited resources and a hacker's aesthetic—a desire to make it smart and simple, using widely available hardware, so that others could follow in his footsteps.

Inside that block of plastic, the Oculus Rift contained a 1280-by-800-pixel screen, split in two in order to show each eye a different 640-by-800-pixel image. Users looked at the screen through a pair of glass lenses. And on the outside, a motion-tracking sensor kept track of which way they were looking.

It was inexpensive. It used an LCD screen—a technology that, thanks to the mobile phone boom, was suddenly cheap and easily available, and displayed bright, high-contrast pictures.

It was responsive. Three decades of progress building microchips and software to power fast-paced video games meant that the system had enough speed to assure low latency, the time from when motion happens to when the view actually changes. That meant less lag and less nausea.

It was small and light. Other headsets bulged with complex systems of heavy optics, designed to magnify images and correct for visual distortion caused in the process. But the Rift used software to pre-distort the image, so it looked perfect after it passed through just two lightweight magnifying lenses.

And it had a wide field of view—perhaps its most important innovation. Before the Rift, the best consumer-level headset on the market had a 45-degree horizontal field of view, about half the range of human vision. Wearing it was like staring through a toilet paper tube. But the Rift's field of view was a stunning 90 degrees horizontal, 110 degrees vertical. It filled the user's entire field of vision, cutting them off from the real world and immersing them in the virtual.

"Palmer was solving a problem, not pitching a marketing group, so he did the job right instead of putting the least important things first," says David Wiernicki, president of Force Dynamics, a Trumansburg, New York, company that builds motion simulators for applications, including driving games. "When you wear a Rift, everything else disappears. Every other display—especially current-generation head-mounted displays—has a frame. You can always see the edge; you always have a reference telling you where the game world ends. With the Rift, the game world doesn't end, because you're in it. People talk about 3-D—hell with the 3-D! What matters here, what makes the Rift a product that outperforms the most frantic hyperbole, is that it eliminates the display device. There's no longer a perceptible display, and when you can't tell that there's a display, your brain says: You're here. And so you're there."

———

Word of Luckey's big breakthrough spread quickly. Virtual reality was still something of a taboo subject in tech circles, and new products were generally met with extreme skepticism—everyone remembered how in the 1990s the technology was over-hyped, then underdelivered. But John Carmack had credibility and wasn't some shady marketer. The few people who were lucky enough to try his Doom 3 demo could confirm that he was onto something.

As a result, a small number of investors and enthusiasts started courting Palmer Luckey for a chance to get involved with his project. One of them was Brendan Iribe, then chief product officer at game-streaming company Gaikai. Iribe had grown up playing video games, and at thirty-two was already an industry veteran: in 2004 he cofounded a company called Scaleform that made development tools for game companies, and in 2011 he'd sold it to multinational software giant Autodesk for $36 million.

"I got a phone call in June 2012 that I needed to meet this guy Palmer Luckey, he's onto something awesome, it's virtual reality and it's finally gonna work," Iribe says. "I thought, 'Nah, VR's never gonna work.' I'm too busy, I'm sorry. Then I looked on the Internet and saw some interesting news coming out of E3, and I thought okay, might as well go . . . I wouldn't want to be the guy that misses the thing that changed the world."

Iribe made contact with Luckey and arranged to meet him for dinner at STK, an upscale steakhouse in Los Angeles. "In walks Palmer Luckey in an Atari T-shirt, shorts, and these awesome flip-flops," he says. "My first thought is, 'Wow, he's a lot younger than I expected' . . . and a slight concern over whether [the restaurant] would actually seat us."

The restaurant was accommodating, and the group—which included Iribe's friends and fellow Scaleform veterans Michael

Antonov and Nate Mitchell—got to talking. "We had this incredible evening," Iribe says, "just listening to Palmer talk about the future of VR, his headset collection, and his infectious vision for where this was all gonna go."

Iribe was still skeptical, but Luckey had caught his attention. A few weeks later, on July 4, the group got together again at a Long Beach hotel for a demo. Luckey arrived carrying a plastic tub loaded with dangling wires, circuit boards, a computer, and the taped-together Rift prototype. He put it all together, turned it on, and handed Iribe the headset.

"When you looked through those lenses, you saw this bold new world," Iribe says. The demo was simple, just a 3-D rendering of a bare-bones room with a short set of stairs leading out of it. But the experience was immersive. Iribe could turn his head and look around, and the Rift responded perfectly.

"You saw this incredible virtual world that was believable," Iribe says. "It was a hint of it, it felt like it. It wasn't all there, but it was enough that it got us excited. We were hooked."

Iribe initially gave Luckey $5,000 to help him prepare for the Kickstarter campaign. But after the demo sunk in, Iribe pushed Luckey to expand his vision. Oculus shouldn't just raise money to sell VR kits to hobbyists, he argued, it should be a full-fledged consumer electronics company. With enough research and development, the Rift could be a mass-market product.

Luckey wasn't convinced. "I didn't know if it was going to be something that I'd earn a living from," he says. "I wanted to sell a few hundred to game developers so that we could experiment . . . I didn't expect it would pick up fast enough for there to be a big industry around it."

Still, outside investment could help Luckey improve the quality of the Rift headset, and fund better marketing materials for the Kickstarter campaign, like professionally produced videos. So Iribe gave Luckey a few hundred thousand dollars of seed capital, and the two men split ownership of a new venture. Mitchell and Antonov made smaller investments and received stock. Oculus VR was born.

———

The company's first order of business was launching the Rift's Kickstarter campaign. Luckey had decided to use the crowdfunding platform because he figured the Rift's core market of VR enthusiasts and software developers would be most willing to take a risk on the unproven device. Under the terms of the fundraiser, anyone who paid at least $300 would receive their own Rift prototype kit, which they could use however they saw fit.

With Kickstarter, you don't have people who are looking to make a large financial return; you have people who just want the thing you're making," Luckey said. If everything went as planned, those new users would start making software for the Oculus platform, building an ecosystem of games and applications that would eventually run on a mass-market consumer version of the device.

Iribe was convinced it would work and pushed to set the campaign's fundraising goal at $500,000. But Luckey worried that was too much. He knew there would be demand from VR enthusiasts, but he wasn't convinced that community was big enough to fund such a large goal. Under Kickstarter's terms of service, projects that fail to meet their target get nothing; if the Rift didn't have the commercial appeal Iribe predicted, the project could die on the vine. So shortly before the fundraiser started, Luckey slashed the target to a relatively modest $250,000, and hoped for the best.

On August 1, 2012, the company launched its campaign with a slick presentation on the Kickstarter website titled "Oculus Rift: Step Into the Game." An embedded video at the top of the page featured Luckey explaining how he started the project, and clips of him tinkering in his parents' garage. It showed off computer renderings of the planned next-generation Rift prototype—a glossy, futuristic-looking headset, no duct tape or random wires. It included in-game footage from John Carmack's Doom 3 demo. And short testimonials from video game luminaries including Carmack, Valve Corporation cofounder Gabe Newell, and Epic Games design director Cliff Bleszinski.

Once the public saw the Rift in action, they turned out in droves. The Kickstarter raised its $250,000 funding goal in less than two hours—and kept on growing.

During that first day of the fundraiser, Luckey was in Dallas, Texas, at the annual QuakeCon gaming convention, running demos of the Rift for interested players. "We were probably the smallest booth at the whole show," Luckey said. "We didn't have any signage, just a black table. And we had a line that was over two hours long the entire weekend. That's when I realized, 'Oh, man, this is gonna be huge. Ordinary people are interested in virtual reality, not just us crazy sci-fi nerds.'"

As the Kickstarter boomed, it became clear that Oculus VR was going to have legs as a company, according to Luckey, and since he was eager to stay engaged with the Rift's technical development, it was time to add some more experienced management. A few weeks into the campaign, Brendan Iribe signed on as the company's CEO. Not long after, John Carmack left a post at Zeni-Max Media to become chief technical officer.

When the Kickstarter campaign concluded thirty days later, Oculus had raised $2,437,429—almost ten times its goal—from 9,522 backers.

The result exceeded everyone's expectations. "I expected more skepticism. I expected that people would need to see it before they believed it," Luckey says. "But even though VR's popularity kind of died down a little bit over the past decade, I think people never stopped wanting it.

"Other products have this problem where they have to convince people what their product does, what they can do with it, and why they should buy it. In our case, people have always known what they can do with VR—movies and pop culture have been selling it to them for decades. All we had to do was step in with a good solution. There's no doubt that the demand was there, just sleeping."

Chapter 5

TWO BILLION REASONS

S ome months after Oculus VR closed its $2.5 million fund-raising campaign, at a cocktail party during a video game industry conference, a well-known game designer—who was fairly drunk at the time, and thus shall remain nameless—poked me in the chest with his beer bottle and demanded to know if I'd "tried out the Rift yet." When I told him I hadn't, his eyes went wide and his voice went up a few dozen decibels.

"Dude . . . *dude!*" he shouted. "You *have* to try it. It's amazing!"

"Yeah, I want to," I replied as I brushed spilt lager from my shirt. "I've heard a lot of hype about it."

"Not hype! This is the real thing. VR is all I want to do now. It's changed everything."

In late 2012 and early 2013, tech circles were buzzing with a new enthusiasm for virtual reality. In January, at the annual Consumer Electronics Show in Las Vegas, Oculus didn't even have an official presence at the convention and still stole the show; simply running demos for press and executives out of a hotel room was enough to get the Rift on several "Best of CES" lists. A *PC Magazine* article called it "virtual reality's greatest hope."

And the excitement wasn't just limited to tech insiders. The successful Kickstarter thrust the Oculus Rift under a spotlight, making Palmer Luckey and his hacked-together headset into stars. The crowdfunding campaign proved there was a surprising level of interest in what most people assumed was a dead technology, but it also put a fresh face on VR. A genius teenager in a T-shirt and sandals—you couldn't make a better avatar for the early twenty-first-century tech industry if you genetically engineered one and incubated it in a lab. In March, at the ultra-cool South by Southwest Conference in Austin, Texas, attendees packed in to see Luckey on a panel discussion entitled "Virtual Reality: The Holy Grail of Gaming." The hottest ticket in town that week wasn't to a film premiere or to see a cool new band—it was a demo of the Oculus Rift.

———

Just about everyone who tried out the Oculus Rift proto-types agreed that they were impressive, even amazing. But the wow factor doesn't always translate into sales. When a consumer version of the Rift headset finally made it to stores, what would keep it from becoming just another novelty for VR hobbyists, engineering geeks, and video game nerds?

When I posed that question to Warren Robinett, a computer game pioneer and VR expert who created games including the seminal Atari 2600 title Adventure, he jokingly suggested a mathematical formula to predict the potential success of the Oculus

Rift: F > $ + K. In plain English: The amount of fun the system provides must exceed the sum of its cost and clunkiness.

"Use as an example the VR entertainment system that Hasbro developed twenty years ago, and then canned without bringing it to market," Robinett said. "Hasbro found that to get a system with a low enough cost that consumers would buy it, the LCDs had to be cheap, the optics had to be cheap, the tracker had to be cheap, and as a result, the VR goggles were clunky as hell. It was pathetic."

In order to succeed, Oculus would have to keep the price of the Rift headset in the ballpark of other consumer electronics—think hundreds, not thousands, of dollars—while still delivering a polished high-end product. Cheap out on any of the components and the result could be literally nauseating.

And the hardware was only half the problem. "Once you pick your design point in the cost versus clunkiness space, you have to give the game designers and programmers a good enough platform that they can actually deliver a decent, fun game," Robinett argued. "And it had better deliver a new experience you can't get on an Xbox or PlayStation . . . like looking down and seeing your guts fall on the ground after you get shot."

In a very real way, the success of the Rift was out of the hands of the people at Oculus; it would depend on whether outside developers produced content consumers wanted. The Rift might be a spectacular piece of hardware, but it would fail without good software. So Oculus didn't waste any time getting out its very first product.

On March 29, 2013, Oculus VR started shipping its first

"development kit," known as the DK1, as a backer reward to more than 7,500 people around the world—half of them outside the United States—who had pledged $300 or more to the Rift's Kickstarter.

"It was largely game developers," Luckey said. "Most game platforms, when they put out development kits, they very strictly limit who they send them out to. They're generally very expensive, tens of thousands of dollars . . . [and] they'll only sell them to certain companies that they approve of. We took the opposite tack. We wanted to make development kits available to everyone, whether they were an indie developer working in their garage, or a triple-A developer with thousands of employees."

Considering that the new headset shipped only six months after the conclusion of the Rift Kickstarter, the campaign's backers received a remarkably polished product. Like the prototype that preceded it, the DK1 had a 1280-by-800-pixel screen and a 110-degree field of view, but it was all built into a custom-molded black plastic shell, not held together with duct tape. The headset came with three different pairs of replaceable lenses, so users could adjust the optics to their vision, or they could twist a knob to move the lenses and screen away from their face, leaving room to wear eyeglasses. The DK1 also had a variety of built-in sensors and trackers—a gyroscope, an accelerometer, and a magnetometer—to help track the user's position and movement. A single wire connected the headset to a control box where users could plug in all the data, video, and power cables that connected the device to their computer. And everything came in a padded, travel-ready hard plastic case decorated with Oculus's corporate logo.

In some ways, the DK1 was obviously still a prototype. The

LCD display was relatively low resolution, so the image it produced was distractingly pixelated. It was also prone to motion blur, so moving objects appeared indistinct and streaky. And it shipped with very little software support or technical infrastructure. It was a device meant for experts to learn from and build on. An ordinary consumer would probably never even get it working.

"We told them straight up, 'We need your feedback. We need you to work with us to make this thing better,'" Luckey said. "I think that really did help us connect with a community of developers who cared about the Rift."

And developers weren't the only people interested in the success of Oculus. Ever since the Kickstarter had concluded, investors had been lining up for a chance to buy into the company.

Initially, Luckey had shied away from outside investment, because he figured an unproven product from an unproven company would automatically be undervalued. If Oculus sold shares, it would have to do so at a deep discount from what the business might eventually be worth.

"VR had never had a successful consumer product, ever," Luckey said. "It had always failed, and there were many examples of companies that crashed and burned, taking hundreds of millions of dollars in investment with them. If I'm an investor, what are the odds that I'm going to want to invest in this new product, no matter how cool it seems, where there's such a precedent for virtual reality failing spectacularly?"

But then the Kickstarter campaign had proved there was real demand for Oculus's technology. And when developers who missed out on the fundraiser turned out in even bigger numbers to order the DK1 directly—eventually the company shipped more than 60,000 DK1 units—Oculus executives decided the time was right to seek outside financing.

"Hardware companies are expensive to run," Luckey said. "It takes a lot of money to design and ship hardware, and you can't afford to try to stretch every penny that you have and not raise money. We had to raise money to keep going."

In June 2013, Oculus closed a $16 million Series A funding round co-led by Spark Capital and Matrix Partners, with participation from Founders Fund and Formation 8—all high-profile Silicon Valley venture capital firms—under terms that set the pre-funding valuation of the company at $30 million. Six months later, powerhouse VC firm Andreessen Horowitz led a $75 million Series B round, with additional capital from all four original firms, at a valuation that put Oculus in the $300 million range.

"The dream of VR had been around so long that most people in the technology community had given up on it," said Chris Dixon, a partner at Andreessen Horowitz. "When we first met Palmer, we saw he not only continued to believe in the dream but also understood how to put all the key underlying technologies together to make it a reality."

Valuing a VR company with a prototype headset at $300 million struck many as euphoric. But just a few months later, Oculus's investors looked like the smartest guys in the Valley.

————

By the end of 2013, Oculus VR was one of the hottest start-ups in the country. Investors were beating at the door, practically begging to give the company money; designers and developers waited for hours at conventions for a ten-minute demonstration of their technology. Sometimes it seemed like every executive in Silicon Valley wanted to schedule a meeting. Most of them got turned away by necessity. But there was one tech icon who was so important that when he called, he got connected instantly.

Mark Zuckerberg, the billionaire founder and CEO of Facebook, was friendly with Brendan Iribe, and they had a few general conversations about how the two companies might work together. Luckey also corresponded with Zuckerberg via email, and the two bonded over shared interests in technology and science fiction. "We started off talking to Zuckerberg because we wanted to show off our stuff," said Luckey. "He's a big fan of virtual reality, and I think he believes in the same vision that we have, that everyone in the world is going to be exposed to VR."

Iribe and Zuckerberg talked several times in early 2014 during visits to each company's offices, and eventually the Facebook CEO had a eureka moment during a demo of the latest Rift prototype. "I'd seen VR before, but this was by far the best experience I'd ever seen," Zuckerberg said. "It was like teleporting to some other place just by putting on a headset, and it was so good that I didn't really want to leave."

Initially, Zuckerberg's conversations with the Oculus cofounders had been general and hypothetical, along the lines of establishing a technical partnership, perhaps a way to adapt Facebook to the Oculus platform. But after his eye-opening demo, Zuckerberg started thinking that Oculus was the key to something bigger—an entirely new communications platform, rather than just a new route to Facebook.

Meanwhile, Iribe and Luckey realized that Facebook's financial backing could actually make VR a mainstream consumer technology. "We really needed to get this hardware off the ground and to consumers, and building hardware is really expensive," Luckey said. "Let's say you want to sell a million of these things—that means you have to have a few hundred million in cash just sitting around to build them." At the time, Facebook had a market capitalization of more than $168 billion, and Zuckerberg had a personal net worth of more than $13 billion, making him the sixty-sixth richest person on the planet. That massive war chest would give Oculus a huge advantage in the unforgiving consumer electronics market.

Once both companies had arrived at the same conclusion, it took less than a week to negotiate the details. On March 25, 2014, Facebook announced that it would acquire Oculus VR in a deal worth $2 billion—$400 million in cash up front, 23.1 million Facebook shares, and another $300 million in performance incentives.

Naturally, Zuckerberg revealed the deal via a post on his Facebook timeline. "Our mission is to make the world more open and connected," he wrote. "We feel we're in a position where we can start focusing on what platforms will come next to enable even more useful, entertaining, and personal experiences."

In Zuckerberg's vision, virtual reality would provide a natural extension of Facebook's services, another way to share your life with friends and family. But instead of posting a picture or video of something after the fact, VR would allow users to actually experience events together, even if they live thousands of miles apart.

"Imagine enjoying a court side seat at a game, studying in a classroom of students and teachers all over the world or consulting with a doctor face-to-face—just by putting on goggles in your home," Zuckerberg wrote. "Imagine sharing not just moments with your friends online, but entire experiences and adventures . . . One day, we believe this kind of immersive, augmented reality will become a part of daily life for billions of people."

In other words, Mark Zuckerberg bought Oculus to build the Metaverse, the collective virtual reality Neal Stephenson imagined in his 1992 novel *Snow Crash*—a place where user-controlled avatars could hang out, do business, and socialize.

It might sound crazy that one of the richest people in the world would spend billions to chase down a sci-fi fantasy. But a few months after the acquisition, I asked Palmer Luckey outright if that was what was happening. "I feel like it's a no-brainer," he said. "The dominating vision of virtual reality in sci-fi is using it for training people, using it for simulations, or using it to access an entire virtual universe existing alongside the real world. That's been a core part of virtual reality science fiction for decades. It's been very clear that it is going to happen, and that at some point there is going to be a virtual metaverse that exists alongside the real world."

———

Facebook's acquisition of Oculus VR was big news. Zuckerberg had bought out other companies before, and had spent much larger sums of money, but his endorsement of a largely abandoned technology made people take notice. If the genius who helped usher in the age of social media believed in the future of VR, maybe it really would be the next big thing.

Businesses started scrambling to figure out how VR might disrupt their industries. David Cole, a veteran of the digital imaging business who had developed VR and 3-D tech for the Eastman Kodak Company, cofounded a start-up called NextVR in 2009 to develop systems for broadcasting live events in virtual reality. Before the Facebook deal, potential content partners had been intrigued but skeptical, telling him "let's wait and see, we want to know what VR is all about, but we're not committing," he said. After Facebook, "the entire world just opened up and changed."

Facebook's commitment legitimized VR and convinced other businesses to start investing in the technology, according to Michael Pachter, a digital media industry expert and analyst for financial services firm Wedbush Securities. "I have been a skeptic of Oculus," he said a few months after the acquisition was announced. "If you don't have a big install base, it's hard [to attract] third-party publishers . . . and if you don't have third-party publishers, you're likely to have very little content. And then Facebook came along. Facebook is looking at the hardware as a platform to do other things—maybe instruction, something like teaching—and Facebook will get third parties to make content. Now Oculus can realize a broader strategy, with a big bank account behind them."

Other analysts remained more skeptical. Some of them told me that VR was unlikely to take off outside of video games and niche entertainment markets like amusement parks. And a few cautioned that the Facebook deal would probably just kick off another fruitless hype cycle, a repeat of the booms and busts of the eighties and nineties.

But many of the loudest critics of the deal turned out to be some of Oculus's earliest supporters—the VR enthusiasts, hackers, and gamers who'd cheered for Palmer Luckey when he was just a kid in a garage with visions of reviving VR through an open-source, community-oriented project. "I feel like I've been waiting in line for an awesome roller coaster, and right when the end of the line is in view, they divert the line to a fucking carousel or something," a former fan wrote in a post reacting to the news on the website Reddit.

Some of the critics assumed that the amazing VR rig they'd hoped for would be dumbed down into a mass-market gimmick designed for chatting with friends and sharing family photos. "I don't want a social platform," another Reddit user lamented. "I want something I can attach to my face to help me forget there are other people." Others worried that Facebook would ruin the Rift with embedded ads and intrusive data collection. And some video game developers expressed concern that Facebook might be an untrustworthy business partner.

A few hours after the acquisition was announced, Markus "Notch" Persson, creator of the blockbuster construction game Minecraft, published a blog post explaining that he had backed the Rift Kickstarter, gotten excited about the technology, and planned to make a version of his game for the Rift. But Facebook's ownership of the technology changed everything. He wrote:

> Facebook is not a company of grass-roots tech enthusiasts. Facebook is not a game tech company. Facebook has a history of caring about building user numbers, and nothing but building user numbers. People have made games for Facebook platforms before, and while it worked great for a while, they were stuck in a very unfortunate position when Facebook eventually changed

the platform to better fit the social experience they were trying to build . . .

I definitely want to be a part of VR, but I will not work with Facebook. Their motives are too unclear and shifting, and they haven't historically been a stable platform. There's nothing about their history that makes me trust them, and that makes them seem creepy to me . . . And I did not chip in ten grand to seed a first investment round to build value for a Facebook acquisition.

Persson immediately called off his version of Minecraft for the Rift, and some smaller game studios promised to suspend their own projects or to cancel pending orders for the development kit. For many early backers of Oculus, the Facebook deal was nothing less than a betrayal, a sellout move, and corporate appropriation of their hobby. They were sure that Zuckerberg and his minions would ruin the product and poison the market. Just like they did in the eighties and nineties, clueless businessmen were going to burst the VR bubble.

Palmer Luckey had expected some of his supporters to object to the Facebook deal. "We knew there would be people who were upset," he told me after the deal went public, "but you can't make decisions based on a vocal minority of people who are going to get pissy. You have to make the decision based on what you think the best thing overall is."

Still, he seemed a little defensive, and perhaps a little hurt by how quickly the denizens of tech-centric Internet forums—a subculture he considered his own—had turned on the Rift. "First of all, the Kickstarter was for the development kit," Luckey said. "There have been other companies that have sold before even shipping their Kickstarter products, and I think that's a real bait and switch. That's not what we did. We never said we're never going to raise money or partner with another company."

The critics, according to Luckey, weren't even real backers of the Rift. "Everyone who's actually buying dev kits, producing games and investing significant portions of their life into VR, almost all of them are very excited because this means there's actually gonna be a platform, it's actually gonna happen," he said. "All the people saying, 'Oh, this is terrible, it's the worst thing,' those people don't really have any investment in virtual reality."

Other critics of Oculus had made significant investments in the technology, and once they saw how much money Facebook spent in the acquisition, they wanted to share in the wealth. In May 2014, John Carmack's former employer, ZeniMax Media, filed a lawsuit against Oculus VR and Palmer Luckey, claiming that Carmack had illegally misappropriated ZeniMax trade secrets when he jumped ship and joined the start-up.

In a complaint filed to the US District Court for the Northern District of Texas, ZeniMax alleged that while Carmack had served as technical director for the company's Dallas subsidiary id Software, he'd spent the company's time and money conducting research into VR, and then inappropriately shared that information with Luckey. ZeniMax also claimed that Luckey had signed and violated a nondisclosure agreement, and Oculus later refused to reasonably compensate the company for its "expertise and know-how."

Without ZeniMax, the company alleged, the Rift would have never been a viable product and Facebook would never have acquired Oculus. "Defendants have wrongfully taken ZeniMax intellectual property and commercially exploited it for their own gain," the complaint said, "and now stand to realize billions of dollars in value."

Carmack denied the charges, arguing that while ZeniMax owned computer code that he had written for the company, it didn't own the insight he'd shared with Luckey. And Oculus's response characterized the lawsuit as little more than a scheme to extort a piece of the $2 billion Facebook deal. "It's unfortunate, but when there's this type of transaction, people come out of the woodwork with ridiculous and absurd claims," a representative said in a statement. "We intend to vigorously defend Oculus and its investors to the fullest extent."

Ultimately, neither the lawsuit nor disgruntled support-ers would stop the Rift's progress. Throughout 2014, top engineering and business talent flocked to join Oculus, including Atman Binstock and Michael Abrash, two pioneering VR engineers who came over from video game giant Valve Corporation; and Jason Rubin, cofounder of the game studio Naughty Dog. "Often they didn't have the intention of working with us until they tried the Rift," said Luckey. "Then they said, 'Oh, wow, this is the future. I have to be a part of this.'"

Oculus also expanded through acquisitions. In June, the company acquired Carbon Design Group, a Seattle product design studio that had worked with Microsoft on projects including the Xbox 360 video game console's controller and Kinect peripheral. In early July, the company bought RakNet, a California firm that made software to power online features in multiplayer games.

Meanwhile, Luckey and his engineers continued to produce new versions of the Rift, and on July 25, Oculus VR started shipping a new prototype headset, the Development Kit 2. Like its predecessor, the $350 DK2 was meant for the use of software developers and researchers, not the general public—Oculus was still a long way from a consumer-ready product. But the updated hardware did represent a major leap forward from the company's first, Kickstarter-funded device.

The biggest change was that the DK2 was a lot less likely to make its users sick. For generations, engineers working on virtual reality had struggled to solve the problem of simulator sickness, the nausea caused by discrepancies between what your body sees and what your body feels. The Kickstarter version of the Rift addressed the issue with built-in gyroscopes (to track when you tilted your head), accelerometers (to track how fast you moved it), and magnetometers (to track which way your head was pointing, as a compass would). All together, these sensors allowed the DK1 to track your gaze in any direction—the device knew when you looked up or down or left or right, and could adjust the image it displayed, near instantly, to match.

But the DK1's onboard sensors tracked only rotational movement; it knew where you were looking but not where you were. If you kept your head steady, looking at a single point in space, and then took a step forward, the system didn't know that you'd moved and didn't update the display. When that happened, your body would feel like you were getting closer to whatever you were looking at, but your eyes would report that the distance to that object hadn't changed. The resulting simulator sickness could end your VR experience in unpleasant ways.

The DK2 addressed this problem by introducing a system to track translational motion, or movement from one place to another. The kit included a small digital camera, and the surface of the new headset was studded with forty infrared light–emitting diodes. Even though the human eye can't see infrared, a camera can, and when the Rift was in use the LEDs functioned like little beacons, telling the computer exactly where the user was and when they moved. Combined with data from the rotational sensors, this new positional tracking system meant DK2 users could look around and move around in the physical world, and their view of the virtual world would always match up. That meant fewer discrepancies between seeing and feeling—and fewer game-over moments of simulation sickness.

The DK2 also reduced the potential for nausea by using a better screen. Whereas the DK1 had used a comparatively low-resolution liquid crystal display, the DK2 featured a high-definition 1920-by-1080-pixel organic LED panel. The upgrade meant that the new screen had a much higher refresh rate—it could update the picture more times each second, so it was less likely to look blurry when a user moved their head. The new display also offered brighter colors, sharper contrast, and better detail. The difference was something like upgrading from a 1990s-era cathode ray tube television to a twenty-first-century flat-screen TV. With the new display, virtual worlds looked more real.

Once Oculus's developers got the new headset in their hands, they started taking it apart to see how it worked and discovered how the company had managed the upgrade. The screen was a repurposed component from one of the most popular smartphones on the market, the Samsung Galaxy Note 3, with Samsung logo, earpiece speaker, and other components still intact. Since the DK2 was just an interim prototype, one step on the long march to a consumer product, Oculus had planned to ship around only 50,000 units, so it didn't make sense to spend the time and money to build a custom display. Meanwhile, Samsung had years of experience making LED screens, and at the time, it was manufacturing millions of Note 3 phones every month, so Oculus was able to fulfill a small order for high-quality components at an unusually low price.

The deal was the first product of an ongoing partnership between Oculus VR and the Samsung Group. Brendan Iribe had first connected with the South Korean consumer electronics giant a year earlier, at a meeting set up by one of Oculus's investors, where he'd expected to pitch Samsung on an arrangement to buy their displays. Instead, he was surprised to find that he was the one being pitched.

"I started talking about what we were looking for, and they said, 'Actually, we've been doing R&D into VR as well . . . we really believe there's an opportunity in mobile and we'd like to partner with *you*."

The Samsung executive showed Iribe a prototype they'd been working on: a 3-D-printed plastic frame with two small lenses, designed to attach to one of the company's mobile phones. Basically, the device was a twenty-first-century update of the Holmes stereoscope, where the pictures were displayed on a phone instead on paper cards.

"I was pretty skeptical at first," Iribe said. "My feedback was, 'This isn't very good. How long have you been working on this?' They said, 'About four months, ever since we saw the Rift at CES.'"

Samsung's execs had recognized that low-cost VR headsets could be a new source of revenue for the company, as well as help drive sales of their mobile phones, but the company's engineers were already stumped by technical problems, like how to achieve precision head tracking. So Iribe suggested that the motion sensor on the phone wasn't good enough, and that the device would work better if it included rotational sensors from the Rift DK1.

Each company put a team on the problem, and a year and a half later, on September 3, 2014, they announced the results of their collaboration: the Samsung Gear VR Innovator Edition, a $199 portable VR headset that used a Galaxy Note 3 smartphone as its brains and screen.

"We expect that there are going to be different categories of VR," Iribe explained. "The highest fidelity experience will run off a computer, but mobile VR offers an ease of use that's going to be incredibly popular." Smartphones were already a ubiquitous technology, and desktop PCs with enough processing power to run high-resolution 3-D simulations were decidedly less common, so Oculus was hedging its bets. The new gadget couldn't compete with the Rift in terms of comfort or image quality, but it was cheaper and easier to use.

Like the first two Rift development kits, the Innovator Edition was marketed as "an early-access, beta-version of the device for developers and enthusiasts rather than a final consumer product." At the time, there were an estimated 19 million software developers around the world, and around half of them were making apps for mobile devices. Oculus and Samsung hoped the mobile headset would inspire that huge workforce to start working on VR, and create the games and programs consumers would demand when the Rift and Gear were finally released to the mass market.

———

Appropriately, the first people to actually receive a Gear were nearly a thousand die-hard VR enthusiasts who traveled from around the world to Oculus Connect, the company's first developer conference, about three weeks after the Gear was released. The event was held at a hotel in Hollywood, California, just steps from tourist attractions like the Walk of Fame and Grauman's Chinese Theatre, but while I was there, I got the sense that most attendees never left the hotel. They were too excited to talk about and learn about VR, packing into seminars like "Introduction to Audio in VR" and "The Human Visual System and the Rift," forming impromptu follow-up discussions in every hallway, and congregating around laptops hooked up to development kit headsets in order to share and get feedback on their personal VR projects.

Oculus, of course, was showing off their own technology, including the latest Rift prototype, code-named Crescent Bay. The company's in-house content team had created a ten-minute collection of "Crescent Bay Experiences"—the same demos that I described in the introduction to this book—that seemed to have a profound effect on everyone who put on the headset. When I finally recovered from my own turn with the prototype, I came back to the demo room and watched for a while as other attendees tried it out. They'd disappear into the cubicles and emerge sporting either a dazed expression, like their mind had been blown, or an irrepressible ear-to-ear grin. When some of my colleagues in the technology press called it "breathtaking" and "a new high-water mark in virtual reality," I started to wonder if we had unwittingly become part of history, like the reporters who attended the first public demonstration of television in 1928—witnesses to the birth of a new medium.

The question now, it seemed, was whether Palmer Luckey would be the new David Sarnoff—the executive who popularized television and led his company, RCA, to become one of the biggest on the planet—or Philo Farnsworth, the inventor of television, who did pioneering work but ended up largely forgotten. When I asked him the question directly, Luckey waved off the comparison. "Instead of 'Hey, here's this wonder kid who started this great company,' I'd much rather see a hundred virtual reality companies selling a billion headsets," he told me.

So far, if success had gone to Luckey's head, he wasn't showing it, and he still seemed to be a typically laid-back—if unusually geeky—California boy. He lived with seven friends (all but one of them Oculus employees) in a shared house in Silicon Valley he unironically called the Commune, where they spent hours playing massive multiplayer tournaments of video games like Super Smash Bros. His one concession to newfound wealth, a newly purchased Tesla Model S, shared driving time with a 1986 GMC conversion van with red shag carpet. "I figure you can buy a Tesla and not be too snooty," he says, shrugging.

The Commune is located in the tony suburb of Atherton, California, just one neighborhood away from Facebook's headquarters in Menlo Park, and Luckey was splitting his time between that office and Oculus's headquarters in Irvine, an hour south of Los Angeles. Facebook had given its new acquisition a great deal of autonomy, mostly providing administrative support and financing, in order to leave Luckey and his team free to concentrate on perfecting the Rift headset. When he found himself in the Facebook office, he avoided boardrooms and meetings, preferring to sit with the engineers and tinker in the hardware workshops. "There's a lot of people here who know all about Oculus but don't know who I am," he said.

Mostly, though, Luckey spent the fall of 2014 on the road. Ostensibly, he was doing research and recruiting engineers—"Actually managing people is a talent that I don't really have, but I'm good at getting people to work with me," he said—but he was also using his newfound fame to evangelize virtual reality. On October 16, he appeared at the National Portrait Gallery in Washington, DC, to receive an American Ingenuity Award from the Smithsonian Institution, and gave a speech to the glamorous gala crowd wearing jeans, sandals, and an Oculus T-shirt under a suit jacket.

At *Forbes* magazine's first annual 30 Under 30 Summit, held in Philadelphia a few days later, Luckey appeared onstage with me to talk about VR, and was received like a rock star by the crowd of 1,500 twentysomething entrepreneurs. Between applause breaks he worked the audience like an old pro, cracking jokes and playacting kung fu moves. When I asked him about the complaints from Kickstarter backers who felt burnt by the Facebook acquisition, he pivoted to the stupidity of governmental overregulation, pointing out how the 2012 JOBS Act prevented start-ups from giving out equity to crowdfunding backers. "Fuck the man, right?" he concluded as the audience roared. Moments later, rather than exit the stage via the stairs, he jumped off the front of the five-foot platform like an excited kid.

Afterward, a steady stream of admirers stopped him to talk as he headed out of the convention center, passing by celebrities like the rapper Wiz Khalifa, and several asked to take selfies with him so they could brag about the meeting to their friends. Luckey was gracious with each, asking questions about their work and giving out his personal email so they could continue their conversations. One fan asked him about favorite video games, and Luckey invited the guy over to the Commune to play them. When he finally got out of the building, an admirer even followed us as we walked back to our hotel, peppering him with technical questions. Luckey paid close attention and answered as best as he could—until he was distracted by a sign in the window of a fast-food restaurant.

"McDonald's Monopoly game is back!" he gushed. "I love McDonald's. I've been fighting this war for years trying to convince people McDonald's is good. It turns out that McDonald's is actually one of the healthiest fast-food places out there. I'm not talking about their salad options, I'm talking about their actual burgers, they're some of the healthiest you can get. It turns out that white bread and greasy meat and sauces are just high calorie. That's just how it is. But the crazy thing is that normal burgers and fries have way more calories, way more fat than McDonald's, and it's because McDonald's actually has reasonable portion sizes. If you go to any diner-type place for a burger, it's so much worse for you than a typical McDonald's meal. That's why I get upset when people are like, 'Fries and burgers are so unhealthy.' That's not McDonald's fault, it's your fault for wanting to eat food that is inherently unhealthy. McDonald's beef is 100 percent beef with no fillers. The only additives are salt and pepper. It is not high-quality beef, it is lean beef, in little shitty patties, that's grilled all the way through, well done. It's not a great patty, but that doesn't mean that it's bad or not meat. There's no fillers or vegetable proteins or other things. The whole pink slime thing never was a problem for McDonald's, and even if it was, they haven't used it since 2007."

As Luckey paused to take a breath, his admirer interrupted to say he was going to try to find his way back to the convention center. Luckey apologized for not being able to solve all his technical problems, gave him his personal email address, and promised to answer any follow-up questions.

It's highly unusual for an executive at a multibillion-dollar start-up to offer a stranger that kind of access, but the gesture wasn't unusual for Luckey. Later, David Wiernicki, president of the motion simulator company Force Dynamics, told me that he'd had a similar experience working with Oculus. "It boggles my mind that [Luckey] was answering emails from some tiny company off in the middle of nowhere," Wiernicki said. "And during the middle of it all, when I asked if there was anyone else I could bother with niggling technical questions instead of him, he replied that as far as he was concerned, his engineers' time was far more valuable than his, and he preferred to take the time himself rather than putting the load on them. For a guy in his early twenties who's been slingshot from a bedroom to a multibillion-dollar CEO in a few dozen months, that kind of perspective is impressive."

On January 5, 2015, a story I wrote about Palmer Luckey was featured on the cover of *Forbes*. Aside from raising his profile even higher—no less than Jay-Z, Puff Daddy, and Bruno Mars have written songs about appearing on the cover of that magazine—it also revealed the extent to which Facebook had made him rich.

After a close analysis, *Forbes* estimated that Luckey owned around 25 percent of Oculus VR when it was acquired, making the deal a half-billion-dollar score; just nine months later, his net worth had already appreciated closer to $600 million, thanks to Facebook's soaring stock. As far as the wealth experts at the magazine could determine, no entrepreneur in history—not even Mark Zuckerberg—had ever made a fortune that big, that young.

———

Of course, it was worth remembering that other young en-trepreneurs had been in Luckey's position before—like Jaron Lanier after he founded VPL Research. "Recent events are weird for me in a way that the folks at Oculus couldn't possibly know, in that so many of the designs, the headlines, the little intrigues, and the chatter are similar to what I experienced over thirty years ago," Lanier told me when I was working on the *Forbes* story. "All I can say is that I wish Mr. Luckey all the success in the world."

VPL filed for bankruptcy two years before Palmer Luckey was born, but he knew its history and knew the obstacles that faced the Oculus Rift. "VR has never had a successful consumer product," Luckey said. "It has always failed, and there are many examples of companies that crashed and burned, taking hundreds of millions of dollars in investment with them."

But he still believed success was inevitable. The time was right, the tech was ready, and Oculus VR was in the right place to deliver. "I don't think there are any comparable products," Luckey said. "We've got the best team in the virtual reality industry, we've got a lot of the best people in the games industry, we've got big partnerships with hardware manufacturers and game developers. I think we're on the path of making the world's best virtual reality hardware . . . and I think the consumer product is going to be way ahead of anything anyone else can do for the next couple of years."

At the rate Oculus was growing, by the time a competing product did catch up to the Rift, virtual reality may have already gone mainstream. But even if it didn't, Palmer Luckey wasn't likely to give up on his dream. "Five years from now, I don't know if everyone's going to have a headset," he said. "But I'll be doing this until it happens, or until I die."

Chapter 6

TAKING HOLD

I n the spring of 2015, my email in-box contained more fantastic ideas about virtual reality than a box full of old cyberpunk novels—dozens of messages every day from public relations firms trying to convince reporters to write about their new clients in the burgeoning VR business. There were pitches for new video game studios ("the company's initial efforts will be focused on creating VR games that everyone can enjoy"), new services ("we provide individuals and organizations with affordable 360-degree video solutions"), and new content ("I'm reaching out to invite you to be among the first to engage with Mountain Dew's brand-new virtual reality experience!"). It seemed like every day I'd receive a new "BREAKING NEWS!" alert from some PR flak announcing yet another Kickstarter campaign for a "REVOLUTIONARY!" and "GROUNDBREAK-ING!" new virtual reality headset—most of which seemed to be backed by fly-by-night companies touting products that would never and probably should never make it to market. One prototype VR helmet promised to deliver a "revolutionary multisensory" experience by blowing hot air onto the users' face, squirting them with water mist, and exuding an assortment of bottled scents—an innovation meant to bring the senses of touch and smell into VR, but which I suspected would just leave its users with wet, stinky faces. (A Kickstarter campaign for the FEELREAL VR mask eventually went unfunded, with only eighty-six backers pledging less than half of the company's $50,000 target.)

I read most of the emails once and deleted them, but a few pitches did catch my attention. One came from a PR firm savvy enough to know the real trick to attracting a reporter's attention. After trading a few emails with a media relations specialist for the Patrón Spirits Company, I found myself in their Manhattan offices surrounded by samples of free liquor. In an effort to market their tequila as a premium handmade product, Patrón had outfitted a specialized drone with seven video cameras, flown the rig through its distillery in Jalisco, Mexico, and stitched the footage together to create a "360-degree video virtual reality experience." After I put on a Rift headset and flew for three minutes through a hacienda full of workers cutting agave, roasting and fermenting the milled *piñas,* and labeling and boxing the finished tequila, I had to admit to the PR team that it was an incredibly effective commercial. I hadn't just been fed a slogan about how the company's booze was handmade, I actually witnessed the bespoke production process firsthand, so I knew it to be true.

"We've always joked if we could transport people to the hacienda, we would never have to spend money on another ad," one of the company's marketing executives told me. Patrón planned to show off the virtual tour at tasting events, he said, and to outfit members of its sales team with Gear VR headsets. "When they're in formal meetings with the head of food and beverage for a large retailer, instead of pitching, 'Hey! Patrón is quality made, blah, blah, blah,' we can just show them our process. At some of our events we're even playing with the idea of, what if we could have the smell of cooked agave in the room while you're experiencing this?"

I said he might want to take a cue from a recently failed Kickstarter and simply squirt tequila in the user's face as they watched the video, but the suggestion was politely dismissed as impractical.

Another unsolicited email pitch caught my attention because of the start-up's A-list backers—which included Google Ventures, the billion-dollar investment arm of the worldwide technology giant—and also because it promised the kind of VR I'd dreamed of ever since I'd read William Gibson's *Neuromancer*. "In AltspaceVR, you can create shared, social experiences with anybody around the world, doing anything on the Internet," the email said. "Want to watch Netflix with your friend in China, go online shopping with your mom in Florida or play chess with a stranger in Israel? With AltspaceVR, you can, and it feels as if they're right there next to you." I didn't have any friends living in China, my mom was in Virginia, and I was terrible at chess, but that still sounded great.

About a week later I met a small team from AltspaceVR at a luxury hotel in Manhattan. When I arrived, they had already taken over a corner of the lobby bar with a massive laptop, a tangle of wires, and a modified Oculus Rift development kit. After quick introductions, one of the company's executives, Bruce Wooden, jumped straight to fitting the Rift on my head, followed by stereo headphones with a built-in microphone. Before I was completely cut off from the real world in this cocoon of plastic, he explained that I could reach out and use a mouse on the cocktail table to move around, and that the headset was outfitted with a sensor device made by a company called Leap Motion that could track the motion of my hands, "so I could just wave or point at stuff." As he pulled the Rift down over my eyes, he offered the barest of instructions: "Hold on a second, and then Eric will be in there to tell you what's going on."

I was in the dark for a moment, and then I was standing on a balcony overlooking a Japanese rock garden. "Hey, Dave," someone said off to my left. I turned toward the source of the sound and saw a tall white robot, its eyes glowing green in a featureless face, its body long and tapered and seeming to hover above the hardwood floor. "I'm Eric," he said, waving a disembodied hand.

Eric Romo, CEO and founder of AltspaceVR, was in his office in Redwood City, California, and I was in a hotel bar in New York, New York, more than 2,500 miles away. But at that moment we were both also in a three-dimensional digital environment created for social interaction in virtual reality. Romo called it Altspace, but it might as well have been the Matrix or the Metaverse. Both of us were running software on Internet-connected computers that allowed us to meet in a digital space, something like a chat room or a multiplayer video game, but without the immersion-breaking presence of a monitor on a desk.

As Romo showed me around, his robot avatar seemed to come to life, his head nodding and moving as he talked, and I could follow his gaze and see where he was looking. His hands could point to different objects in the environment or simply gesture for emphasis as we conversed. Even though the simulated environment looked like a 3-D video game and the robot bodies were highly abstract, it felt like we were actually in the same room; through gaze and gesture, his digital avatar drew my attention, and the normal rules of social interaction kicked in. I kept eye contact, I nodded when he talked, and as a result, I felt a kind of human connection that's never present in a text conversation or on the phone. At one point when I moved my own avatar inexpertly and found myself apparently inches from Eric's face, I automatically jumped back and apologized to him; it felt like I was violating his personal space.

For about half an hour, Romo demonstrated some of Altspace's different environments—a movie theater, a living room, a backyard with a fire pit, a corporate conference room. Altspace was designed to import web applications as 3-D objects, so he opened up web pages and video streams, and they appeared as giant TV screens floating in the virtual space. I could move toward a video and the volume would get louder, and then it would diminish as I moved farther away. The same thing happened when we entered a room full of other Altspace users, who were congregating in several small groups; I found I could move in and out of individual conversations, just as if I were attending a social gathering in the real world.

When the demo was over and Eric thanked me for my time, I said, "It was nice meeting you," and without thinking about it, reached out to shake his hand. His avatar did the same, and our virtual appendages briefly intersected in space. It seemed silly, but the gesture had come naturally. I felt as though I'd actually made a personal connection; the meeting felt real, even though I knew that it wasn't. And when I removed the headset, I had the uncanny feeling of being displaced. One second I was having a chat in a garden with an interesting new friend, and then suddenly I was sitting in the corner of an upscale bar while confused Manhattanites gave me horrified looks over their champagne and blood orange liqueur cocktails.

Later that day I returned to my office and found an email from Altspace's PR manager waiting in my inbox. "As a journalist I realize that family and friends may not always understand exactly what it is you do," it read. "The next time this confusion comes up, please send them the attached pictures. I think they will clear everything up."

The photos were of me sitting on a couch in the hotel lobby, wearing an Oculus Rift headset and large over-the-ear headphones, seemingly cut off from the world. My head is turned away from the camera, and I'm gazing at something that is not there. I am oblivious to the fact that Bruce has moved to sit next to me on the couch and, with a huge grin, is reaching around me and holding up two fingers behind my head.

That was the day I learned the hidden cost of immersive virtual reality—the constant threat of bunny ears.

———

As start-ups like AltspaceVR scrambled to produce vir-tual reality applications and content, some of the world's biggest tech companies were jockeying to beat the Rift to market and establish a hardware standard.

Japanese consumer electronics giant Sony was one of the first to officially throw down the gauntlet. In March of 2014—just a week before Facebook announced its plans to acquire Oculus—Sony revealed it was in the early stages of developing its own VR headset, code-named Project Morpheus, which would function as an accessory for its PlayStation 4 video game console. "We see it first and foremost as another way of building vibrancy and value into the PlayStation ecosystem," Andrew House, then president and CEO of Sony Computer Entertainment, told me a few months later. "It's unusual for us to come out and talk about a product this early in the process, but by opening up to the development community and not doing everything behind closed doors, it's got a much better chance of succeeding. And the wave of interest we're getting is incredible."

Google entered the fray in June of 2014 when it simultaneously announced and released a surprisingly low-tech VR solution at the end of a keynote address during its annual developers conference: a kit made of corrugated cardboard with two built-in lenses, which could be folded and taped to form a box just big enough to hold a smartphone. The device, called Google Cardboard, began its development only six months earlier as one of the company's "20% time" projects, where employees are encouraged to pursue interests outside of their day job. The idea's origins, however, probably dated back to 2012, when USC professor Mark Bolas released a similar DIY kit called the FOV2GO viewer.

Google didn't intend for Cardboard to be a stand-alone product; the design was released to the world under an open-source license, free for anyone to produce, distribute, and either give away or sell for a couple of dollars. Instead, Google hoped the viewer would serve as a platform to other products, like smartphones running its Android operating system, or as a VR interface for applications like YouTube and Google Maps. The announcement also served, according to some tech industry observers, as a way to mock Facebook's $2 billion acquisition of Oculus. It seemed like Google founders Sergey Brin and Larry Page were telling their Silicon Valley neighbor Mark Zuckerberg: "Good luck selling $300 headsets. We're going to give ours away."

On March 1, 2015, Taiwanese consumer electronics company HTC unveiled another high-end headset, the Vive, the result of a collaboration with video game and software developer Valve Corporation, based in Bellevue, Washington. "We believe that virtual reality will totally transform the way that we interact with the world," then HTC chief executive Peter Chou said at an event announcing the device. "It will become a mainstream experience for general consumers . . . the possibilities are limitless."

At the time, HTC was mostly known as a smartphone manufacturer, but was expanding its product line to include portable electronics like fitness trackers and digital cameras. Meanwhile, Valve was one of the video game industry's most respected software companies—its very first product, the 1998 first-person shooter Half-Life, is widely regarded as one of the most influential games ever—and Steam, its digital distribution platform for PC gaming, dominated the industry's digital sales channel since it was launched in 2003 as a platform for the company's own titles.

———

Valve had worked on virtual reality for years, and in many ways was directly responsible for the growth and success of the Oculus Rift. The project that would become the Vive began in earnest in 2011, after programmer Michael Abrash—an alumnus of companies including Microsoft and id Software—joined Valve and began working with a team building wearable displays. By 2012, the group had developed an ungainly prototype that amounted to an LCD screen attached to head-mounted digital cameras. The device kept track of its user's orientation and movement by scanning the room, identifying a series of visual markers (basically large square bar codes) mounted on the walls, and working out where the device was located in relationship to these landmarks. Valve's engineers loved the way the system allowed for "room-scale" VR, where users could walk around their physical space and have that movement reflected in the virtual world. But they weren't eager to go all in on a system that required consumers to post weird bar codes all over their room.

So Valve kept investigating the technology, and encouraged other companies to come up with better solutions and supported some of the most promising projects. When Oculus's Kickstarter campaign launched in August 2012, a widely distributed promotional video for the Rift featured endorsements from both Abrash ("I'm really looking forward to getting a chance to program with it and see what we can do") and CEO Gabe Newell ("It looks incredibly exciting, if anybody's going to tackle this set of hard problems, we think that Palmer's going to do it"). Newell's recommendation, in particular, lent instant credibility to the project: a headline on the video game news site Kotaku proclaimed that "John Carmack and Gabe Newell Are Very Excited About This Virtual Reality Gaming Headset."

Meanwhile, Valve continued to develop its own hardware and software. By January 2013, the company had developed a monocular prototype called the Telescope. When users looked through a lens and moved the device around, it made it appear as if they were gazing into a computer-generated world. In March, Valve released a free VR Mode update to its game Team Fortress 2, allowing the multiplayer first-person shooter to run on virtual reality headsets, including the upcoming Rift development kit. And in April, the company's engineers completed their first prototype for a low-persistence, head-mounted display with built-in positional tracking—an oversize and ugly piece of hardware that was never meant to leave the lab, but represented a big first step toward a commercially viable virtual reality headset.

Valve's team continued to tinker over the spring and summer, and by fall they had something they were ready to show off—a new system that combined optical tracking, motion sensors, high-quality optics, and low-latency displays into a demo called the Room. Starting in September, engineers and developers from around the tech world made the trek to Bellevue for a tour of eighteen virtual spaces ranging from urban street scenes to the surface of Mars. Inside Valve's room-scale VR environment, the simulations seemed especially immersive and real; since users could move around the actual room, it made them feel present in the virtual one. The demos played to the strengths of the system—users had to walk around obstacles, duck under pipes, and dodge swinging mechanical arms. Witnesses called the experience "absolutely incredible," "world changing," and "lightyears ahead of the original Oculus Dev Kit."

That last claim was of particular concern to one early visitor to the Room—Oculus VR CEO Brendan Iribe. In September 2013, the Rift Kickstarter had just ended, and the Oculus management team was working on finalizing their prototypes and getting the development kit out the door. But when Michael Abrash called Iribe and invited him to visit Valve to try out their demo, he dropped what he was doing and jumped on a plane to Washington.

"Abrash called and said, 'Brendan, we have something I think you're going to want to see,'" Iribe said. "When Abrash calls and says that, you go."

In Bellevue, Abrash and Valve lead VR engineer Atman Binstock led Iribe to a small room where the walls were covered with square bar codes—the fiducial markers that allowed the system to track its location in space. As they entered, Abrash made an off-hand comment to Iribe.

"[He] said, 'You know, no one's gotten sick yet, I'm curious to see what you think,'" Iribe remembers. "I said, 'Not even you?' He said, 'No.' Michael and I are the most sensitive guys out there. Literally a few head turns and we're out. So I was intrigued but very skeptical. Because it had still been hard for me to experience VR for long periods of time—or even short periods of time, to be honest." Abrash strapped Iribe into a clunky 3-D-printed headset with exposed circuit boards and dangling wires, and then started the demo.

"Then came this game-changing moment, a moment that I will absolutely never forget, when I knew VR was really going to work," Iribe said. "It was going to work for more than just enthusiasts and nerds like us. It was going to work for the entire world.

As I looked around, I felt great, and I felt like I was there. All of a sudden, the switch in the back of my head flipped. Instead of thinking, 'Wow, this is a really neat VR demo' I was in it, and I believed I was there . . . that's the magic of presence. I hadn't felt it before until that moment, and it felt great."

Iribe left the room with a new sense of purpose. "The bar had been set, this was it," he said. "This was what we had to deliver for the consumer VR to really work. You had to feel like you were there comfortably. You had to get presence."

He also left knowing that while Oculus was far closer to a consumer-ready product, Valve's technology offered a better experience, and that he'd need their help in order to catch up. So over the following months, the two companies kept talking and sharing their knowledge. Valve set up one of their demo spaces, known as the Valve Room, at Oculus headquarters. In January 2014, the two companies announced that they would collaborate to "drive PC VR forward," and Valve said it had no plans to release its own VR hardware. In March, at Valve's own developers conference, Abrash even urged the company's developers to check out the Rift and said Oculus was the company most likely to ship a viable VR headset within the next two years.

Then Facebook started talking to Brendan Iribe about an acquisition, and communication between Oculus and Valve came to a sudden halt. On March 11, Atman Binstock left Valve to lead a new Oculus R&D team based in Seattle. On March 25, Facebook went public with the details of the $2 billion deal. And three days later, Michael Abrash left Valve to work as chief scientist at Oculus.

Now that it had access to Facebook's wallet, Oculus could poach whatever talent they needed, build their own teams, and throw money at whatever technical problems popped up between the release of the first Rift development kit and an eventual consumer version of the product. But Valve now found itself without key personnel or a hardware partner. Its years of research had produced impressive technology, but the company had no clear route to commercializing what they'd discovered.

And that's where HTC came in. The Taiwanese electronics company was looking for ways to expand its business beyond smartphones, and its Future Development Lab had tinkered with virtual reality hardware enough to realize the technology could be the next big thing. So the company reached out to Valve, and sometime in the spring of 2014, HTC cofounder Cher Wang sat down with Gabe Newell and worked out a deal.

———

Less than a year later, on March 5, 2015, just a few days after HTC and Valve officially announced their virtual reality headset, I got to try the Vive at the annual Game Developers Conference (GDC) in San Francisco, California. While Oculus had dominated the news in the period following the announcement of the Facebook deal, HTC and Valve had been busy building a high-end VR system to compete with the Rift—and now thousands of professional video game designers were practically beating down the doors of the showroom to get their first look at it.

My demo took place in a small room constructed in the south hall of the Moscone Center, a corner of the convention complex hidden away from the public somewhere deep under Howard Street, behind multiple security checks and a phalanx of public relations agents. The space was cold, heavily air-conditioned so that the heat produced by the high-end gaming PC in a corner didn't turn it into a sauna. A tangle of cords ran from the PC to a collection of strange objects made from plastic, glass, and nylon. After I entered, an attendant started picking through the inventory and strapping pieces on my body: the large black Vive headset, its surface dotted with indentations that had sensors at the center; two SteamVR controllers, each the size of a TV remote but with fewer buttons and a round touch pad. All of this was connected to a harness that snapped around my waist and tethered cables going to my head, my hands, and the PC in the corner. When the attendant pulled the headset down over my eyes, I was momentarily in darkness. I felt the cold air in the room, heard the hum of the PC, gripped the controls, and felt like a space monkey ready to blast off into orbit.

When HTC announced the Vive, they billed the device as the first VR headset to offer a full room-scale experience—a headset that let users "get up, walk around and explore your virtual space, inspect objects from every angle and truly interact with your surroundings." The hardware was similar to what Valve originally developed for the Room, but where that system used a camera on the headset to see its position relative to markers on the walls, the Vive's tracking system, code-named Lighthouse, turned the idea inside out.

Instead of cameras, the system used projectors—two little boxes attached to walls in the demo room, each of which contained an array of LEDs and two scanning lasers. In order to track the location of the Vive hardware, the Lighthouse boxes first emitted a quick burst of infrared light, and then the lasers swept across the room, one from side to side and one from floor to ceiling. When tiny sensors on the Vive headset detected the infrared flash, they started counting the microseconds until they saw the beam from the lasers. Then after all the lights had been registered, the computer knew exactly how much time it took for each source to reach each sensor, so it could calculate the exact distances involved and create a model of the system. Repeat this process sixty times a second, and the Lighthouse could track a user's precise position, orientation, and movement.

And because the Lighthouse flooded an entire room with light, it could track more than just a headset. Each of the controllers terminated in a chunky hexagon of black plastic that was studded with the same IR sensors embedded in the headset. As a result, users could control a simulation either by pushing buttons on each controller or by simply pointing and gesturing. The system would always know where its users' hands were and what they were doing with them.

The value of the Lighthouse system was immediately apparent from the beginning of my Vive demonstration. It started in a simple virtual room, a sort of blandly high-tech space that could have been torn from any of a dozen cyberpunk novels and movies. But even though the room itself was uninteresting, what was inside it instantly grabbed my attention: two virtual versions of the SteamVR controller floating right where my body's own proprioceptors told me they should be. For the first time, I could feel physical objects in my hands and see them in virtual reality.

When I moved my arms, the virtual controllers moved with me. Then I moved my right thumb to press down on one of the controllers' touch pads, and something unexpected happened: a tiny red balloon inflated out of the end of the device and then, with

a pop, floated off into the air in front of me. I knew it was a simulated object, but I had created it with my own hands, so I reached
forward to touch it, out of instinct—and batted it away off the end
of the controller. I pressed the button again, made another balloon,
and then punched that one, bouncing it off into the distance.

It was a simple thing, but the effect was profound: I'd felt
present in VR before but had never felt so physically connected. I
could touch and manipulate virtual objects the same way I did in
the real world—with my hands, with movement. The controllers
bridged realities and turned me from a passive observer into an
active participant.

Another part of the demo drew me in even further. When it
started, I found myself standing in a dark, featureless space. The
controllers were still in front of me but had changed appearance
slightly. The one in my right hand now had a point on the end, like
a stylus, and the left displayed a small color wheel, like you might
see in a computer painting program or an art textbook.

I knew what to do. I pointed the stylus at the wheel, clicked a
button, and selected a color. Then, squeezing the trigger, I waved
my hand through the air and painted a bright floating ribbon. The
room was my canvas, and my hands were the palette and brushes.
I drew circles, then a smiling face, then wrote my name in the air,
all in different colors. Before long, I'd surrounded myself with
little virtual sketches, and I stood at the center of a gallery of my
own making, its walls made of paint, all just an arm's length away.

As I admired my handiwork, I heard a disembodied voice
break in from outside the simulation. The HTC attendant who was
running the demo had been watching my progress on a computer

monitor. "Why don't you walk a few steps straight ahead," he said, "and then take a look back at what you've created?"

In my excitement over the SteamVR controllers, I'd forgotten the real point of the Lighthouse system was to allow users to move around in virtual reality; I didn't have to stay in one place to do all my painting. So I stepped forward, walked through the curtain of painted colors and then into the dark, featureless space where I'd started. After a few paces I turned around and looked back at my creation. A dome of glowing graffiti sat there in the nothingness, centered on the point where I'd been standing.

Now that I realized I could move around, I started experimenting with different objects in the virtual space. I drew a bright yellow orb and then, next to it, a smaller black one; when I walked around them, it looked like the moon eclipsing the sun. I drew a long line across the room, as straight as I could, and then looked at it from the endpoint; the narrow strip of paint almost disappeared in my view, as if I were peering down the length of a wire.

As I scrambled around the virtual space, looking around, above, and below my creations, I forgot that I was actually in a small demo room in a convention center. At one point, while jumping though a virtual hoop, I would have crashed right into a wall if I hadn't inadvertently triggered a safety mechanism called the Chaperone system—a glowing grid, like a virtual fence, that appears in the virtual world as you approach the safe boundaries of your simulation. When the Chaperone faded up into view, I stopped and let out a small gasp—in part because I remembered where I was, but also because the bright grid against the dark space looked an awful lot like the inside of a *Star Trek* holodeck.

———

The Vive put Oculus on notice. The Rift's high-profile
Kickstarter campaign had won the company lots of news cover-
age, and Palmer Luckey's kid-genius act kept buzz going through
the long months when Oculus had nothing new to show. But now
an established global consumer electronics company had teamed
up with one of the brightest stars in the game business and stolen
the spotlight. Tech reporters covering GDC called the new demo
"spectacular" and "the best ever made," and software developers—
the content creators critical to the success of any new computing
platform—were lining up to get their hands on a kit. Even worse
for Oculus, HTC seemed ready to deliver right away. When CEO
Peter Chou announced the product, he promised that a developer
version of the Vive would be available within a few months, and
that the company would actually ship a final consumer version by
the end of 2015.

In contrast, Oculus had already produced two development
kits, but refused to say how many more prototypes it had planned
or when it might actually release a consumer product. No one
thought the Rift would hit the shelves in 2015. It increasingly
looked like the "first practical VR headset" wouldn't actually be
the first to market.

In fact, it might not even be the second. Sony was also putting
the heat on at GDC. On the same day I had my Vive demo, I attended a press conference where Shuhei Yoshida, president of
Worldwide Studios at Sony Computer Entertainment, announced
that the Project Morpheus virtual reality headset would launch in
the first half of 2016. The company even had a few prototype units

on hand for attendees to try out, and while the demos—most notably, a virtual trip to the deep sea in a shark cage, which concluded with a disturbingly realistic great white attack—didn't impress me as much as what I'd seen on the Vive or the Rift, I was very impressed that the demo was running on a $399 PlayStation 4 video game console, instead of a high-end gaming PC that probably cost several thousand dollars. While Oculus and HTC fought over the relatively small market of high-end PC gaming enthusiasts, Sony planned to go mainstream and sell a slightly less powerful headset that ran on a device 5.6 million gamers around the world already had in their homes.

In the two and a half years since virtual reality had reentered the public consciousness, Oculus VR had seemed to be the clear leader—maybe even the only serious competitor—in the race to finally deliver a consumer-ready VR product. But in the days after GDC, it seemed like the company was being eclipsed. Many of the developers I spoke to at the show said they were more comfortable working with a company like Valve than with Facebook. Others figured they'd make more money designing VR games for the ubiquitous PlayStation console than they would for a select few enthusiasts who owned cutting-edge gaming PCs.

Oculus attempted to calm some of those fears on May 6, when it revealed new details about the first consumer version of the Rift, dubbed CV1, in a series of blog posts on the company's website. First, the company promised, it would start taking preorders for the headset later in 2015, and would actually ship the device sometime in the first quarter of 2016. Second, the company released some of the technical specifications of the device, in an attempt to prove that a brand-new computer capable of running a

Rift could be purchased for as little as $800—twice as much as a PS4, but far less than the several thousand dollars I'd imagined.

Just over a month later, on June 11, 2015, Oculus VR officially unveiled the final consumer version of the Oculus Rift at a small press event in San Francisco. Brendan Iribe kicked off the event by showing off the final form of the headset, a slim fabric-covered device that included two high-definition OLED screens, built-in headphones, and a single external tracking sensor.

"We're able to finally deliver on the dream of virtual reality," Iribe said. "This is going to change everything . . . We've all been dreaming about this for decades, and it's finally here."

Iribe didn't say how much Oculus would charge for that dream or when, exactly, it would ship the product. But the press event did give him a chance to show off a slick-looking headset and quell some of the fears that the Rift had fallen behind in the market. And after Iribe debuted the hardware, Palmer Luckey took the stage for a surprise announcement—Oculus Touch, a pair of handheld controllers that tracked a user's hand movements, just like the Vive's SteamVR controllers.

After the event was over, Luckey remained on the stage and put on the Rift headset. Print and video news photographers swarmed onstage as the twenty-two-year-old whiz kid looked toward the sky, as if gazing at virtual objects that only he could see inside the (unplugged) gadget. Then he raised his hands, showing off the Touch controllers, and pretended to be boxing a virtual opponent. The crowd surrounded him, squeezing together as tight as the front rows of a rock concert, taking pictures and shouting questions.

Two months later, Palmer Luckey appeared on the cover of

Time magazine in almost the same pose, wearing the Rift headset, alongside the text: "The Surprising Joy of Virtual Reality . . . And Why It's About to Change the World."

———

A few months later, at Oculus's annual developers confer-ence in Hollywood, I interviewed Luckey for the first time since the CV1 was announced. When I entered he room, he bounded over, shook my hand, and then just as quickly retreated and threw himself prone onto a couch.

"Busy day, huh?" I asked. "How's it going?"

"I'm really running on fumes," he said.

Hearing this, an Oculus PR representative who was working the meeting stepped forward. "Palmer, what can I get you, man?"

"You cannot help with what ails me," Luckey said, sitting up. "I am exhausted in body and mind."

"Aside from preparing for this conference, what have you been working on?" I asked.

"Touch has been the biggest project. I'm actually focused on a lot of things— No, that's kind of oxymoronic . . . I'm distracted by a lot of things. Back in the early days when it was just me, I had to work on everything, and now I don't have to, we have smart people working on everything. So what I do is get distracted and try to work on everything, because I care about everything that's going on."

"Are you still doing any hands-on engineering? Did you do any on the Touch controllers?"

"Oh, definitely. I'm doing top-level stuff too, and I do a bunch of PR stuff like we're doing right now, where I trick people into repeating things that I say. What a great trick. It's like you talk to them and then they go out and they're like, 'Hey, here's what Palmer said.' How often does that happen in real life? But other than that, I still do real work. I think of myself as a hardware guy. I'm really passionate about making our hardware better."

"Is there a particular engineering problem that's really engaged you over the last year?"

"The ergonomics around Touch were a big challenge, making a controller that worked well for everyone and still felt natural. When you're making a tool, like a screwdriver, you can make compromises. It might be a little too big for one guy, a little too small for another guy, but that's fine as long as it works, because you only use a screwdriver for a couple minutes. But with Touch, you might use it for hours and hours, and it has to work for me with my gigantic hands, all the way down to somebody with little bitty hands. I think it's been one of the most difficult ergonomic challenges ever, because you have to make something that works for everybody, and you want them to feel like it's actually their real hand. That's pretty rough."

"It's not about fit," I said. "It needs to feel like an extension of you."

"Yes, but even beyond that," he replied, "because an extension might mean holding a tool. You want it to feel like you're not holding anything at all, that you just have hands in the virtual world."

"So is the Touch design locked down now? Are both Rift and Touch ready to manufacture?"

"Well, there's a big difference between having a design that's locked and a design that is actually coming off of a manufacturing line. You have to actually manufacture this thing with real processes, real machinery, real automation—see what the issues are, make tiny little adjustments. And like, when I'm talking about tiny, I don't mean like trimming a few millimeters here and there . . . and then getting all of the notches and the levers and the straps and everything feeling like they're really dialed in. There's a lot of work going from a one-off prototype to something where we're cranking out thousands of them every day."

"And you've had to invent so much of the manufacturing process, right? It's all custom manufacturing, not something you can simply hire a partner to build."

"I'm glad you recognize that, and I can't tell you too much about it, but some of the fixtures we have on our assembly line may be literally the most complex automated manufacturing steps for any consumer product in history. There are more complex [processes] for certain scientific instruments that we're aware of, but for a consumer good, these might literally be the most complex, ever."

"So you're doing a lot of fine-tuning and polishing at this point?" I asked.

"Yeah . . . and over the last few months, I've personally been . . . I won't say 'wasting,' but I've been spending a lot of time on media."

I laughed. "So I'm not wasting your time, but this isn't your favorite part of the job."

"It definitely isn't a waste," Luckey said. "It's very valuable, especially when it comes to VR . . . It is super important [to explain] something that hasn't really come into the consumer consciousness. But at the same time, you're right, it's not my most favorite thing to do."

"How are you handling the attention? The cover of *Forbes*, the cover of *Time*. How do you feel about that?"

"This sounds super douchey, so just bear with me. There are a bazillion guys in business who get this kind of attention. Sometimes it's just their fifteen seconds of fame . . . But what's actually going to make me successful isn't being on a cover as face of the month, it's getting the Rift on there, so people say, 'What is that thing? Why is it special?'"

I sat up in my chair. "You know, I was having lunch recently with my eighty-something-year-old uncle who was visiting New York. I don't see him much. He asked me what I've been working on, and when I told him I was writing a lot about virtual reality, he asked, 'Is that what that kid was doing on the cover of *Time*?'"

"Exactly," Luckey said. "The people who are looking at the covers of magazines or looking at newspapers, they are probably not aware of virtual reality. So making them aware of it is the first step."

"Do you get recognized on the street?" I asked.

"Very rarely. It's because I am a master of disguise," he said, laughing. "I guess my appearance changes somewhat regularly. Like, do you see my hair, how it's cut? I look super, super fresh. I don't look like a hippie. I get recognized at game conventions and tech conferences. But on the street, it's pretty rare."

"Have you gotten advice from Zuckerberg or anybody else in the industry about what's gonna happen?"

"I don't think it's gonna happen to me like it's happened with Zuck," Luckey said. "Zuck is truly the face of Facebook. Right now I'm that for Oculus, but I think that in a year there's gonna be way more people who know about the Rift than about Palmer Luckey. That number's gonna be vanishingly small."

"So you're the story that gets people talking about the Rift, and once the product is out, you fade to the background?"

"I hope that VR is a way bigger story than whatever I'm doing, because my story hasn't changed over the last few years. I'll admit it's a pretty good story. It's like one of those stereotypical 'he came from nothing and look where he is now' stories. But you can only tell it so many times, and then there's plenty of other examples of new people doing good things. It's only interesting for so long."

———

Maybe Palmer Luckey's star would eventually fade, but in the last few months of 2015, VR was getting hotter and hotter. In October, toy industry giant Mattel relaunched its View-Master stereoscope as the View-Master Virtual Reality Viewer, repositioning the seventy-six-year-old novelty as a high-tech gadget. The new viewer, designed in partnership with Google and based on its Cardboard technology, was a $30 plastic case that fit around a smartphone; if users ran a View-Master app and peered through the headset, they could view 360-degree 3-D environments including the Australian Outback, the Tower of London, or the surface of the moon. Additional sights were available for purchase both via download and on $15 "experience reels" reminiscent of the classic View-Master cardboard photo disks.

And virtual reality wasn't just a hot product. It had a certain high-tech cachet that made it a must-have accoutrement to any cool event, premiere, or party. In just one twenty-four-hour period in mid-October 2015, you could go from Fifth Avenue in Manhattan (where apparel company Tommy Hilfiger set up Gear VR units so shoppers could watch its fall runway show from a virtual front-row seat, before buying) to Wall Street (where a black-tie crowd that included model Karlie Kloss and singer-songwriter Usher walked the red carpet into a luxurious charity gala, and then fit headsets over their perfectly coiffed hairdos to take a virtual tour of a run-down schoolhouse in Toklokpo, Ghana) to just off Times Square (where attendees at the premiere of the horror film *Paranormal Activity: The Ghost Dimension* could demo a new VR game based on the film franchise).

VR was also booming in the news business, where executives who'd been blindsided by the rise of the Internet seemed determined to get ahead of the next big digital medium. In early November, *The New York Times* distributed more than one million Google Cardboard viewers to its subscribers, bundling the kits in with their weekend newspapers, and released a free smartphone app that allowed users to watch 3-D documentary films and try out interactive features like a virtual tour of New York City.

At a launch party for the app in Manhattan on November 5, members of the media mingled and sipped cocktails while executives gave speeches about the profound impact VR would have on journalism. "*The New York Times* has been telling stories for a hundred and sixty-four years, and this feels like one of those rare moments where the way we tell stories has changed," said Mark Thompson, president and CEO of the New York Times Company. "[It's] about how you try and use every available means to bring the world to people in a more engrossing, more compelling way than before. It's incredibly exciting for us."

The first story available on the app, a documentary titled *The Displaced*, focused on three children uprooted by war and caught in a global refugee crisis. "It's a story we report about all the time . . . but it's also a story that is sometimes difficult for readers to understand and connect with," said *New York Times Magazine* editor in chief Jake Silverstein. By using VR, the *Times* could actually put readers into a war zone, and allow them to see and hear from the victims of war firsthand, building a personal connection to the story.

"There are now sixty million people worldwide who have been displaced by war and persecution, and that's more than at any time since World War II," Silverstein said. "It was very important for our initial efforts . . . to be used to tell a story that we all know is important, that we all know is serious, and that we all want to understand much better.

"So I'm afraid I have to ask you to put your drinks down," he said.

At that, the attendees reached for Cardboard viewers that had been issued to them upon entering the party, each preloaded with an iPhone running the *New York Times* VR app. After a short period of confusion where a handful of ink-stained old newshounds demanded assistance turning on the gadget, the crowd went still and silent—deaf to the world with headphones in their ears, blind to reality because of the boxes in front of their faces.

Instead of joining them, I kept my viewer in my lap and observed the party guests as they watched the ten-minute video.

Their heads turned in apparently random directions, sometimes seeming to stare at a wall or the ceiling, sometimes tracking the motion of objects that weren't really there but must have seemed real to them. A few people's mouths fell open in surprise. Others frowned and looked distressed about what they were watching. It reminded me of the famous photograph from *Life* magazine of the audience at the 1952 premiere of a 3-D movie—a room full of elegant people wearing strange headwear, all lost in a shared illusion. Megaloptic creatures more startling than anything on the screen.

Chapter 7

VR AND CODING
IN LAS VEGAS

We were somewhere around Utah on the edge of the troposphere when the embarrassment began to take hold. I remember thinking something like "I feel stupid wearing this in public; maybe I should take it off." And suddenly there were clowns all around me and the sky was full of shirtless gymnasts wearing sequined pants, swinging and spinning on nylon straps, even though I was in a jet going about four hundred miles an hour to Las Vegas. And an orchestra was playing and a woman was singing something that sounded like French.

Then it was quiet again. I had removed the VR headset and popped out my earphones when I noticed the guy in the window seat looking at me. "What is that?" he asked, gesturing at the brick of black-and-white plastic in my lap. "It's a virtual reality headset," I said. "I was watching a video." I yanked my phone out of the device and shoved everything into my carry-on bag. No point in mentioning the Cirque du Soleil show, I thought. The poor bastard will be in Vegas soon enough.

It was January 2016, and the consumer version of the Samsung Gear VR had been out for two months. Most stores were still sold out. Very soon, I knew, wearing a headset in public might be completely normal behavior. But today it was still strange and made me self-conscious. I would have to ride it out. The annual Consumer Electronics Show was about to get started, and when I got there I'd be trying dozens of new virtual reality products. More than 170,000 people from around the world were expected to attend the tech expo, filling the convention center to capacity and spilling over into hotel ballrooms down the Vegas Strip . . . and I was, after all, a professional journalist; so I had an obligation to cover the story, geeky or not.

———

It's hard to resist pretending to be Hunter S. Thompson when you're a reporter chasing a story to Las Vegas. Then you realize you're on an air-conditioned monorail to gadget central, not a savage journey to the heart of the American dream. The Consumer Electronics Show takes over the city. The billboards promote Chinese contract manufacturing companies instead of casino stage shows, and when a stranger hands you a flyer, it's as likely to say "visit us at booth 79450" as "girls, girls, girls."

Since 1967, the Consumer Technology Association's annual trade show had become the go-to event for companies trying to sell their new products to retailers and the media. In previous years, exhibitors debuted groundbreaking consumer technologies, including videocassette recorders, camcorders, compact disc players, and high-definition TVs. This year, VR was the must-see technology.

Official CES marketing materials breathlessly explained that because of "the fast growth of virtual reality," organizers had increased the square footage of the Gaming and Virtual Reality Marketplace by 77 percent over the year before, and more than forty exhibitors would be there, including Oculus, Sony, and HTC. Before the show even started, news outlets like *The Guardian* and *USA Today* were already proclaiming CES 2016 "the year of VR."

Even members of the world's oldest profession were in on the hype. A few days before the show began, I received a press release from Sheri's Ranch, one of Nevada's legal brothels, announcing that while they enjoyed an annual surge in business from CES attendees, management was "bracing against" the growth of virtual reality because the convenience of simulated sexual satisfaction "threatens the Ranch's customer flow." Fortunately, the release went on to explain, "courtesan" Red Diamonds remained confident that "virtual reality couldn't teach a man what I could teach him."

Of course, CES hype doesn't always bear fruit. In 1995, *Newsweek* magazine wrote that "virtual reality is finally coming to the living room" after trying four different headsets at the show; unfortunately, the most successful of them was Nintendo's ill-fated Virtual Boy, and that was discontinued in the United States after less than six months. Then there was the CES I'd attended back in 2009, when 3-D televisions were supposed to, yes, finally come into the living room: I have fond memories of Sony chief executive Howard Stringer's keynote address and his bringing actor Tom Hanks onstage to show off the company's wonky-looking plastic 3-D glasses. Hanks wisely kept himself above the hype with a naughty-child routine, reading marketing text off the teleprompter in a comical monotone and sniping at the product. "Oh, look, they're so cool and hip . . . I think these are the best glasses Sony's ever made," he deadpanned. By 2015, just 9 percent of US households owned TVs with 3-D capability, and by January 2017 all the major manufacturers had dropped the feature from their new models.

But even if virtual reality didn't have a winning record in Vegas, at least this time around it looked like a high roller. My first stop at the show was a gilded ballroom at the Wynn Las Vegas luxury resort, where HTC and Valve were laying down their cards and showing off a new prototype of their codeveloped Vive system. (Convention-goers who couldn't get past the velvet rope were left to fight for a space in line at the Vive Truck, a massive eighteen-wheeler outfitted with eight demo bays that spent four days parked near the convention center.)

The second (and final) Vive development kit, called the Vive Pre, was never supposed to exist. Soon after HTC announced the headset at GDC, they indicated that a commercial version would ship in the fourth quarter of 2015—an ambitious timeline that would give the device a nice head start on the Oculus Rift. But then months passed with no official launch date. Devotees started to worry that something was wrong with the system. On December 8, their fears seemed to be confirmed. In a news post on Facebook, HTC announced that it would start shipping a new development kit in January 2016, and then follow up with a commercial version in April—a few days after the Rift hit the market. The comments on the post read as if HTC had canceled Christmas—"very sad and confused"; "I fear the Vive project has seriously run off the rails"; "words cannot contain my disappointment."

A week later, HTC CEO Cher Wang tried to put out the fire by promising that the delay was due to "a very, very big technological breakthrough"—an innovation so big that releasing the first Vive without it would be unfair to early adopters. "We shouldn't make our users swap their systems later just so we could meet the December shipping date," she said. But standing in a Vegas ballroom, holding the Pre in my hands, I wasn't buying her explanation. The new headset was smaller and lighter, but aside from that, the only change I could see was a small round lens embedded in the front of the device.

At that point, I *might* have made a sarcastic comment to the HTC employee assigned to give me a demo about how adding a camera was such a small thing it couldn't possibly be worth a six-month delay. He responded by demonstrating both how the new addition works and why I am an idiot.

Here's an obvious but important truth about virtual reality. In order to convince someone that they're present in a virtual world, you have to separate them from the real one. Fortunately, the limitations of our current technology actually help achieve this. We need to put an LED screen in front of the user's eyes, and that keeps the user from seeing what's in front of them. The screen and the tracking hardware are relatively big and heavy, so we need to enclose them in a bulky headset, and that blocks the user's peripheral vision. Headphones deliver convincing stereo sound, but they also have to sit on or in a user's ears, silencing the world around them.

All this isolation makes it a whole lot easier to convince a VR user that he or she has left the real world and gone elsewhere. But it also means that the illusion of presence depends on maintaining isolation. I have found this works something like biological night vision. The longer I'm in a VR environment, the more I grow used to it and forget about where I really am . . . but if I have to pop the headset off, even for a second, when I return to the VR, it seems artificial again. My tendency to perceive the virtual as real can be regained only through another prolonged exposure.

Think about how many times while reading this you've glanced from the page, maybe to check the time or have a sip of water. Now imagine if every time you did that, you had to go back to the beginning of this section or chapter and start over. That's like what happens when you have to take off your virtual reality headset. It makes it hard to get back where you were.

So what are we supposed to do when we want to take a drink?

Enter what I like to call the Vive's "where's my beer?" button. When you're inside any virtual environment, you can double-press a button on the Vive's controller, and that unassuming new camera lens on the front of the Pre headset records the world in front of you and sends it to the computer for processing. The Vive then displays, in bright blue outlines, a live view of the real world presented as a virtual one; it looks something like the neon-highlighted world of *Tron*, or perhaps the scene at the end of *The Matrix* where we see Neo's point of view of a world made of glowing code.

The practical result of all this is that it lets you see the world around you without taking off your headset. But the brilliance of the system is that it's not simply giving you a peephole—it's mediating between the virtual and physical worlds, displaying your real environment as if it were an artificial one. You can interact with it, but it doesn't remove you from the VR experience. Essentially, you don't lose your night vision.

This isn't the first time someone has put a camera in a VR

headset. Oculus flirted with the idea in some of its Rift proto-types, and Samsung's Gear VR has a "passthrough camera" mode you can turn on to look through your phone's rear lens—perhaps because you're using it on a plane and want to know whether the guy in the window seat is giving you the stink eye. But using that feature just dumps you out of whatever you're doing and turns on a video feed. It doesn't do anything to protect you from the shock.

The Pre's camera also helps to address a problem specific to VR systems like the Vive that are built for walk-around, room-scale experiences. What happens when a user steps outside of the play area? Forget about breaking immersion—wandering players risk breaking their neck if they walk into a wall or trip over an unseen obstacle.

The first Vive prototype I tried out, back at GDC, addressed this problem with the Chaperone system that reminded me of a *Star Trek* holodeck. As you approached the edge of the play area, it faded in a wall-like grid in front of you, warning you not to pro-ceed. Now, with the Pre's front-facing camera, the Chaperone sys-tem went even further. Come close to a border and the same grid appears, but beyond it, there's a fainter, less detailed version of the "where's my beer?" Neo-vision. The virtual world doesn't disap-pear, but in the distance, it intersects with the physical world, and faint lines trace out the edges of real-world objects.

The practical result of this is that you're less likely to stumble over furniture. But it also turned a limitation—the Vive's finite play area—into an interesting feature. In the new Chaperone sys-tem, the grid isn't a barrier, it's a transition. Step into it, and your Neo-vision activates, allowing the user to perceive a space where the world is both real and virtual.

To me it suggested a future where VR headsets like the Vive and the Rift merge with smart eyewear like Google Glass and Microsoft HoloLens, allowing people to live in multiple realities at once, and shift from one to the other as needed. I imagined being in my office, but my trans-reality glasses hiding the cubicle farm and making it appear to me as if I'm on a beach in the Caribbean. My desk is there too, though the live video feed of it appears to be sitting on sand, not cheap carpet. And if my boss approaches, the system quickly fades out the virtual beach, and lets me see the entire real-world office.

———

It was a nice thought, but for the time being, I'd have to be content with staying home and pretending to work in a virtual office. After checking out the Chaperone system, I played a new game in development for the Vive, Owlchemy Labs' Job Simulator. It's a tongue-in-cheek farce set in the year 2050, after robots have taken over all human labor; the premise is that a particularly sympathetic robot created a simulator so bored humans could learn what it was once like "to job," and relive the glory days of work by pretending to be a gourmet chef, office drone, convenience store clerk, or automotive repairman.

It turns out that the robots got it wrong. When I put the Pre headset on and started the game demo, I found myself standing in the middle of a small cubicle at the center of a bright cartoon-style office. The cubicle was covered with standard bits of office detritus, but there was something off about almost all of it; the computer keyboard had only two keys, 0 and 1, and my coffee mug was printed with the legend *World's Most Average Worker*.

I was admiring the *Make Job Happen* motivational poster on the back wall of the cubicle ("Why job tomorrow what you can job today?") when my supervisor—a floating computer monitor displaying a happy face and wearing a dangling red tie—arrived to give me the day's first assignment. "Workers would traditionally start their day with an addictive liquid stimulant," he announced in a mechanical voice. "Time to get the java flowing!"

I turned back to the desk and, in the real world, raised my hands, which were holding the Vive's motion-tracked controllers; at the same time, a disembodied pair of white gloves rose to float in front of me in the cubicle. HTC had made the controllers smaller and sleeker since the last time I used them, but their function was the same; I reached toward my coffee mug, squeezed a trigger, and "gripped" it, locking the mug to my hand in the game and to my movements in the real world. Then I turned around, stuck the cup under the spout of a small coffee machine, and poked a big red button on the machine with the gloved fingers of my other hand. The cup filled with animated liquid, and without thinking, I knew how to finish the assignment—lifting the cup up to my mouth and tipping it toward me.

It must have been an odd sight for the HTC employee running me through the demo—a guy with a plastic lunch box strapped to his face trying to drink from a joystick. In the game, the coffee disappeared from the mug, and the supervisor spoke again. "Also, workers would ingest a frosted sugar torus for sustenance."

You can see where this is going. Job Simulator parodies the working world by distilling it into a series of banal, pointless actions, and the player instinctively and mechanically completes them. I was about ten tasks in and inserting a virtual CD-ROM disk into a virtual computer so I could make a virtual presentation for my virtual boss when I realized the irony. It's a "what the hell am I doing?" moment that marks a turning point in the game, and seems to usually elicit the same response from players: destructive rebellion. I spent the next few minutes picking up every loose object in the cubicle and either throwing it across the office or smashing it on the floor. The violence peaked when I realized that if I picked up a stapler and shook my hand, I could make it fire staples at my supervisor like a semiautomatic pistol.

It took barely ten minutes in Job Simulator for a simulated office job to become a blur of impotent rage and failed assignments. Maybe the robots actually got it right.

———

After the Vive event, I went to bed early, and the next morning I woke up before dawn. I walked up an eerily quiet Las Vegas Strip to an empty CES press room and connected my laptop to one of the few available hardwired connections to the Internet. Then I waited, drank coffee, and compulsively hit the "refresh" button on my browser. It was January 6, 2016, and at eight A.M. Oculus VR's online store was finally going to start taking preorders for the first consumer version of the Oculus Rift. This was a moment I'd been anticipating since the company's Kickstarter campaign closed three years earlier, and I wasn't going to miss placing one of the first orders because I was stuck on dodgy hotel Wi-Fi.

Oculus had announced the presale in a blog post on the company's website two days earlier but hadn't divulged many details aside from the place (Oculus.com) and the time (8:00 A.M. on

January 6, 2016) where overanxious VR nerds could give the company our money. We still didn't know what would come in the box, when it would ship, or what it would cost. But as I browsed the Internet while waiting for the store to open, I could see that a horde of people around the world were also counting the seconds to the presale and ready to order, whatever the price.

Naturally, there was plenty of speculation. The Rift development kits sold for $350, but those headsets were nowhere as advanced or polished as the first consumer version (or CV1). At the Oculus Connect conference in September, Palmer Luckey had said that the CV1 would cost more but be priced "roughly in that ballpark." Did that mean $400? No one was sure. The Rift was the first entry in a whole new category of computer hardware, and since Oculus invented many of the CV1's components and developed new manufacturing techniques, we couldn't even make an accurate estimate of the company's costs. One Goldman Sachs forecast estimated the Rift's bill of materials cost—the total price for all the components, materials, and supplies required to construct one finished product—was $500, and that didn't include significant expenses like marketing and software development. Another analyst told me each unit cost Oculus $600. There was no way to be sure.

Whatever it was, it seemed likely the CV1 would be sold at or below cost. One of the unusual things about the video game industry is that new consoles like the PlayStation or Xbox are considered loss leaders and priced as low as possible. A company like Microsoft will gladly take a $100 loss on each unit, because they don't see that transaction as selling hardware—they're buying real estate in your living room. Once they're in the house, *then* they can make money, on everything from licensing deals to subscription fees to advertising and game sales. There's no doubt that executives at Oculus (and the people holding the purse strings at Facebook) saw the Rift the same way. Get it out, get it in as many living rooms, workplaces, and engineering labs as possible, and do it before the competition starts staking their own claims.

At eight A.M. Oculus answered the big question with a new post on its website. "We're excited to announce that Rift is now available to preorder for $599," the post read, "and it will ship to 20 countries/regions starting March 28." Every headset would ship in a custom carrying case along with a tracking sensor, an Xbox One controller, and a new peripheral, the Oculus Remote—basically a six-button TV remote control. And every purchase would be bundled with two free games, the cartoon-styled platform adventure Lucky's Tale, and the multiplayer space dogfighting simulator EVE: Valkyrie.

I didn't even finish reading the post before I started clicking through the online store. Shopping cart: Oculus Rift, Qty: 1. I clicked "checkout," filled out my shipping address, clicked "continue." At this point the surge of customers pounding Oculus's website caused the next page to slow for a second while loading, and I noticed a little voice in the back of my head. "Five ninety-nine?" it said. I ignored it, typed in my credit card information, hit "continue" again. The site hung for a few seconds this time. "Five ninety-nine? That's not in the ballpark of $350," the voice squeaked. "That's barely the same game!" Sorry, voice of caution. Shipping address, payment method, order details, all okay. Shipping: $30.00. Estimated total: $629. Sure, fine. "Agree & Complete Order." Click. Awesome. I felt like a kid who just mailed his annual letter to Santa.

Other consumers—many of them young gamers who don't have much cash, or adults who have more financial sense than I do—weren't so excited when they saw the news. "Hype train just derailed and fell down a cliff," said the top-voted comment in a discussion about the Rift's price on Reddit. "This is what billions in funding from Facebook gets you? Unreasonably high consumer launch price?" wondered another. The anger poured out in comment after comment, from "they gotta be kidding" to "dreams shattered."

Many overseas customers balked at their even higher prices. "The actual price with shipping is €742," one European user pointed out—an amount equivalent to more than eight hundred US dollars. "Are you fucking kidding me, 43 [euros] for shipping. Better be delivered by Palmer Luckey himself."

If you judged by the comments, it was all adding up to disaster. "Honestly I think this is yet another death blow to VR," wrote another user. "At these prices VR will never achieve mainstream approval . . . and will remain little more than an expensive novelty."

Later in the day Luckey logged on to the site and tried to put out the fires. "I share your concerns," he wrote. "Believe me, I want nothing more than for VR to succeed in the long run . . . The unfortunate reality we discovered is that making a VR product good enough to deliver presence and eliminate discomfort was not really feasible [at] lower prices.

"To be perfectly clear, we don't make money on the Rift," he continued. "The core technology in the Rift is the main driver [of cost]—two built-for-VR OLED displays with very high refresh rate and pixel density, a very precise tracking system, mechanical adjustment systems that must be lightweight, durable, and precise, and cutting-edge optics that are more complex to manufacture than many high end [digital camera] lenses. It is expensive, but for the $599 you spend, you get a lot more than spending $599 on pretty much any other consumer electronics device."

The $350 "ballpark" figure, Lucky said, was a miscommunication based on earlier statements that a Rift *and* a PC fast enough to run it would cost around $1,500. "As an explanation, not an excuse: during that time, many outlets were repeating the 'Rift is $1500!' line, and I was frustrated by how many people thought that was the price of the headset itself. My answer was ill-prepared, and mentally, I was contrasting $349 with $1500, not our internal estimate that hovered close to $599 . . . that is why I said it was in roughly the same ballpark."

The explanation seemed to help calm the critics—and from all indications, their cries of doom were wrong anyway, as Oculus was selling Rift headsets faster than they could make them. When I submitted my order a few minutes after eight, the confirmation said my Rift would ship in March 2016; customers checking out just a few minutes later received estimates in April. All day long, the estimated shipping date for new orders kept slipping further into the future—a strong indicator that the Rift had sold out of its initial stock, and that demand exceeded Oculus's manufacturing capacity. By the end of the day, new orders had a ship date in July. Oculus wouldn't say how many sales it made, but several analysts I talked to came up with a consensus estimate of around 70,000 units that first day, and ultimately, more than 400,000 preorders before the March 28 release date.

———

Maybe the critics were right and a $600 price tag would scare off consumers. But it didn't look like that was happening on the CES exhibit floor, where Oculus VR's booth was surrounded by a line of people waiting more than an hour to see the Rift for themselves.

It was a very different scene from Oculus's first appearance at CES. When the company made its inaugural visit in 2013, only a few months had passed since the conclusion of the Rift's Kickstarter campaign, and a small team of employees worked out of a suite at the Venetian hotel, running demos on two pre-DK1 prototype units. The modest effort was still enough to land the Rift on several "Best of CES" lists: *The Verge* wrote that "the Oculus Rift is the most revolutionary gaming experience we've seen in years" and *Wired* said that it was "the coolest thing we've ever put on our face." In 2014, Oculus made its official debut at the show, though its space consisted of private meeting rooms in the ballroom of a hotel north of the convention center, well outside the main action. Still, new demos won more best-of-show recognition. In 2015, the company arrived in the main convention center with a big booth that was open to the public. As the tech officially crossed from underground sensation to mainstream must-see, the "Crescent Bay" prototype earned Oculus even more "Best of CES" honors.

This year, Oculus's booth was one of the biggest at the show—less a booth, really, than a two-level building housing a dozen Rift demo rooms and a lounge space where attendees could sit and play with the Gear VR. Another auxiliary space next door included more demo bays and a few meeting rooms.

I found Palmer Luckey inside and we headed to a conference room to chat. He karate-kicked the door open with a sandaled foot and jumped into a chair.

"You seem pretty energetic," I said. "I thought you might need to recover from arguing with your critics on the Internet."

"I don't need any recovery." He laughed, leaning back on the chair's rear legs. "I'm used to far worse—I used to run a web forum. I'm not gonna claim I'm any kind of Internet master, but I think that most people, in corporate settings particularly, don't understand how to interact with an audience like that. And that's not because I have special training. It's just when you're naturally part of that group, it's a lot easier."

"I talk to a lot of executives who didn't grow up on the Internet; it's sort of an alien thing to them," I said. "They look at people in web forums as the rabble with pitchforks."

Luckey nodded. "It's hard to be angry about something you don't care about. It doesn't make sense to treat people like a mob, when in reality they're a big chunk of your customer base."

I asked him how he felt about the preorder announcement and launch.

"Well, preorders are going really, really well, better than expected," he said.

"Above your projections?"

"We're really happy with how it's going, yeah."

"What about the people who say $600 is too much money?"

"The fact is, that's just what this technology costs right now," Luckey said. "I can see where it's too expensive for some people, but at the same time, it's not outside the realm of lots of other mainstream consumer products." The same people complaining about a $600 VR headset might have a $600 monitor or television in their living room, he argued, but only the headset seems expensive, because VR is a new product.

"When high-definition TVs came out, they cost thousands of dollars," Luckey explained. "People didn't bat an eye, and that's because they understood that of course the prices would dip down over time. There was an expectation there, because TVs had always been expensive when new technology shipped. But people don't have that same expectation with VR. The only thing that sets expectations so far is Google Cardboard, Gear VR, and our two dev kits, which were way less complex. So when the Rift comes out, we're kind of setting a new understanding."

"You're not just establishing the price of a product," I offered, "you're establishing the value of this medium."

"Right," Luckey said. "And it's not like anyone's going to come out with something that's drastically better and undercuts us."

So why choose to make the Rift a high-end product? I wondered. Oculus could have worked on making the Rift cheaper and, as a result, put VR into more people's hands; instead, the company made a specific decision to sell a more expensive device to a much smaller group of people.

"We want everyone who tries it to say that it's awesome," Luckey said. "Even if it's not something that they want yet, or if they're waiting for more content, or they're waiting to get a better PC, or they're waiting for the cost to go down. That's a lot better than people trying VR—on any headset—and saying, 'Oh, I tried it, it isn't very good, I'm not interested.' We kind of have an unprecedented wave of hype right now, and I want to make something that capitalizes on that and puts good VR in the eyes of as many people as possible."

Still, I argued, doesn't a high base price raise the risk that you won't reach enough people? And isn't that an even bigger problem outside of the United States, because of international shipping costs?

Luckey thought for a moment. "Shipping globally is really hard," he said. "We're shipping in about twenty countries right now. We're working to expand that, but all these countries have different regulations and different certification processes. It's really easy to be mad about high prices and say, 'Oh, they must just hate my country.' But it's just really hard to ship all around the world."

"So your message is, 'We don't hate you, Australia, it's just really hard to get stuff to you,'" I joked.

"Australia is a relatively small market for us," Luckey said. "There is a cost to setting up distribution and shipping and doing it to a relatively small number of people. We don't want to abandon Australia or any other country. But we also can't afford to just say, 'Oh, it's really expensive to be in this country? Well, we'll just charge you the same amount.' Apple does that in some countries, but they have a lot of profit margin to spend on making sure their products ship around the world, and they can make different amounts of money in different countries to try to keep things fairly even. But when you sell hardware at cost, you don't have that luxury. When it costs more to ship something, that cost is directly reflected to the consumer, because there's no margin that you can use to absorb it."

As he spoke, Luckey got more animated, the way I'd seen him perk up in the past when talking about a new engineering challenge. "I'm surprised to see you so excited about shipping logistics." I laughed. "It's like this is a puzzle that you have to figure out."

"I am excited," Luckey agreed. "But it's less like a puzzle and more like . . . I guess to use a school example, it's less like a test, and it's more like homework. It's about doing the right paperwork, waiting for the deadlines, filling out more paperwork, meeting with the right people, turning things in, getting things back. It isn't like you just go, 'Bam! We've got it, we're all set to go in twenty countries.'"

"Is there any part of your job or the product launch process that you're not excited about?" I asked. "Something you're happy to give to somebody else?"

"I try to stay involved in pretty much everything," Luckey said. "I can't imagine wanting to hand off anything entirely."

"Are you still involved in engineering work?"

"Yeah, but I split it between a lot of things," he said. "Right now, this week, there's no engineering; it's all business meetings and press meetings. But that's what happens as you get closer to shipping a product. You aren't doing novel engineering for the headset. You're working on things like improving manufacturing, getting units off the line faster. There are people working on next-generation stuff in research, but generally speaking, most of us are focused on getting the product out the door."

"Do you get to work with Michael Abrash or John Carmack anymore?"

"Oh yeah," Luckey said. "I was just visiting Carmack in Dallas two weeks ago."

"What did you see there?"

"Whataburger is really good. I was like, 'Wow, Texas knows how to do burgers.'"

I laughed. "You're from California, you've got In-N-Out Burger. That's one of my big regrets about living on the East Coast, we're never gonna get In-N-Out."

Luckey leaned forward in his chair and pointed at me with his index finger. "You guys have Smashburger. You have Five Guys."

"Have you tried Shake Shack?" I asked.

"Oh my god, Shake Shack is so good. I love the Shake Shack in D.C., that's the only place I'll eat in D.C. I get the Shack Stack, the patty and portobello mushroom with cheese."

"I'd be embarrassed to admit this to other people," I told him, "but I know you'll understand—one of the factors I considered when deciding to move into my current apartment was that there's a Shake Shack two blocks away."

"Oh, that's great."

"I didn't tell my wife that was one of the factors, but it definitely was." I paused, looked down at my notebook, and flipped through a few pages. "What was the next question I was gonna ask? Now all I can think about is burgers."

Luckey laughed. "That's how it gets at conventions, man. Nothing but dry sandwiches in these parts."

Completely distracted by my stomach, I resorted to asking a cliché question directly out of my notes. "So, can you give me an example of something you were surprised to learn about running a business in the last couple of months?"

Luckey hesitated, and just for a moment, his usual high energy and enthusiasm dipped. I might have even detected some fatigue—a perfectly reasonable though uncharacteristic result of years of hard work getting the Rift to this point. "I don't know." He sighed. "Let's wait till we ship, and then I'll talk about the lessons I've learned. We're still in the middle of trying to get this thing out there. It's easy to promise; it's hard to deliver."

"You're real heads-down right now," I said. "Not much time for reflection."

"When you take orders, that's when you really have committed," Luckey said. "We've given a date, March twenty-eighth. That's when we are going to ship these things. The period between when you make that promise and when you deliver is the toughest, 'cause you can't stumble."

"So what happens on March twenty-ninth?" I asked. "Do you get to take a vacation?"

He smiled, energized again. "I'm not taking a vacation, because that would mean not working on VR."

———

After I finished talking to Luckey, I decided to check out some of the other VR start-ups with booths on the CES floor. The Gaming and Virtual Reality Marketplace took up only a tiny portion of the convention's 2.47 million square feet of exhibit space—something the size of a supermarket inside an area bigger than 43 football fields (American football, specifically—equivalent to 32 standard FIFA football pitches, 11 Australian rules football grounds, or 308,750 foosball tables)—but it was packed with more people than any other section of the show. VR was new and exciting, and there was something new around every corner.

Across an aisle from the Oculus booth, a long line of people waited up to three hours for a five-minute demo of the Virtuix Omni, a $699 high-tech treadmill billed as "the first virtual reality interface for moving freely and naturally in your favorite game." The base of the Omni was a five-foot-wide shallow bowl made of low-friction plastic, so if you walked on it while wearing special slippery shoes, your feet slid up and down the sides in a near approximation of natural gait. Motion sensors tracked the action and allowed you to control a video game character by walking or running in place. Add the VR headset of your choice, and the movement helped convince your brain you're striding through virtual worlds, not moonwalking on an oversize dinner plate.

Just like the Oculus Rift, the Omni was initially funded by a crowd of gamers who didn't mind paying in advance for a chance to try it out. Mechanical engineer Jan Goetgeluk started working on the device as a hobby project while he was employed as an investment banker at JPMorgan; after two years of tinkering he quit his job, founded Virtuix, and in June 2013, launched a Kickstarter campaign in order to raise funds to bring the product to market. The campaign passed its initial $150,000 funding goal in just over three hours, and in six weeks raised more than $1.1 million from 3,249 backers. In December, Goetgeluk appeared on the reality television series *Shark Tank* to ask the program's celebrity investors for $2 million in exchange for 10 percent of the company. The sharks didn't bite, but after Facebook shelled out $3 billion for Oculus VR a few months later, billionaire panelist Mark Cuban changed his mind and joined ten other individual and institutional investors to contribute $3 million in funding. Now Virtuix was at CES to show off the Omni's final production model, which it said would start shipping in the third quarter of 2016, about nine months later.

A few feet away, the "Chinese Oculus" was showing off its newest hardware. ANTVR, a start-up born out of a Beijing technology incubator, earned the nickname after it raised more than $260,000 in a June 2014 Kickstarter, but its reputation soured after it shipped its first headset six months later. The All-In-One Universal Virtual Reality Kit could connect to computers, game consoles, or Android devices, but it was poorly constructed, made of cheap materials, and panned by VR enthusiasts. After that the company pivoted to manufacturing low-end Google Cardboard alternatives like the $25 ANTVR TAW, a simple plastic frame for a cell phone connected to an elastic strap.

But ANTVR hadn't given up on high-end virtual reality. Most of the company's booth was roped off with retractable belt barriers, and behind them, a young man wearing a bulky black headset walked an erratic path around the pen, like a hallucinating drunk trying to navigate imagined obstacles. I stood and watched awhile, thinking about how VR turns its viewers into a sight to be viewed, before a staffer came over and explained the man was wearing ANTVR's next-generation Cyclop prototype. Like the Vive, the Cyclop had a built-in camera on the front of the headset, but ANTVR used it to constantly scan a three-foot area around the user's feet, looking for reflective stickers laid out in a grid on the floor. It was a clever way to track a user's movement and deliver room-scale VR without the use of external sensors, though I doubted whether consumers would be willing to stick so much junk to their floors.

As I walked on, I found more mobile VR headsets around every corner. Parisian start-up Homido showed off their $80 competitor to the Samsung Gear VR, as well as a $15 take on Google Cardboard called the Homido Mini, a set of collapsible lenses resembling a lorgnette that snapped onto a phone. A Hong Kong company called I Am Cardboard had a booth full of more familiar $10 cardboard kits, many of them custom-printed with advertisements for clients like Adidas and Lionsgate Films, who used them as promotional items. Carl Zeiss AG, a 170-year-old German optics company best known for high-end camera lenses, was showing off the $120 Zeiss VR One—a bit pricey but highly customizable, since Zeiss encouraged tech-savvy users to use a 3-D printer to create adapters and make the headset compatible with many different models of phones.

Other companies were looking to get rid of headsets entirely. A French start-up called Scale-1 Portal was demonstrating a VR media system called Scalee, which appeared to be little more than a video projector in a desktop-computer-sized box. The device projected a blurry image onto the back wall of the booth, and a man wearing dark glasses stood in front of it, staring; as the image changed, he moved his body from side to side as if he was dodging obstacles. Stereoscopic 3-D, I realized, like in a movie theater, but paired with a motion-tracking camera so users could interact with the video. "It's virtual reality that doesn't require a headset," a promotional model working for the company told me when I stopped to watch.

I continued past booth after booth of companies pushing new VR stuff. Giroptic and 360Heros had new high-definition 360-degree cameras; Vitrio was hustling an app and motorized tripod that let real estate agents create virtual home tours. And the VR craze spilled out of the virtual reality marketplace into every corner of the convention center. Many companies were using VR to sell unrelated products. Audi had a virtual car showroom set up where users could put on a Vive headset and walk around a computer-generated model of its R8 supercar, or even stick their head through the hood to look at the engine. The company said it planned to start installing the demo at dealerships, where it would allow customers to see what a car might look like with different configurations, paint colors, and options. The semiconductor company Qualcomm invited attendees to try out Gear VR headsets loaded with cell phones powered by Qualcomm microchips, a chain of logic that attracted VR-curious attendees to its booth and allowed the company to boast that their processors were the best at delivering mobile virtual reality. Even NASA got in on the action: the space agency's first-ever CES appearance offered convention-goers the chance to put on a Rift and make a virtual visit to Kennedy Space Center or ride on an Orion spacecraft.

VR was also a spectacle—some of the biggest crowds at the show clustered around demos that showed off the intersection of virtual reality and motion simulation technology. In one expo hall, CES attendees gaped at an exhibit that appeared to have flown in from outer space. Moveo, a free-rotation VR simulator built by Krush Technologies, looked like a starship escape pod made of dark glass and shiny white plastic; a person-sized cockpit was mounted on a series of rotating gimbals so that it could spin in any direction. Users put on a helmet that was fitted with an Oculus Rift headset, climbed into the simulator, strapped into a pilot's chair, and used an airplane control stick to navigate around a VR spaceflight simulation. I watched as the simulator spun someone through a barrel roll and then head over tail before I elected to skip a potentially stomach-turning demo.

But right inside the convention center's main lobby there was another VR motion simulation that I couldn't pass up. In the middle of one of CES's most heavily trafficked areas, Samsung had set up a "Gear VR Theater" and invited attendees on a virtual roller-coaster ride: three dozen stadium-style movie theater seats rigged to tilt and shake in unison while the people sitting in them watched a VR recording of a trip on Six Flags Magic Mountain's Twisted Colossus. (You might also know the coaster as the Screemy Meemy, one of the rides the Griswold family hijacks at the end of *National Lampoon's Vacation*.)

I figured moving chairs couldn't affect me too much, that it would just be a novelty, so I strapped in—and was shocked by the reality of the experience. The shaking motion of the seats didn't replicate anything like the forces on an actual coaster, but still, the combination of seeing and feeling motion, along with hearing the people in seats around me scream and cheer at all the right moments, made the VR simulation much more real than I expected. My heart rate soared, my stomach lurched, and I forgot I was at a convention in Las Vegas, not screaming through a zero-g roll in a six-car train on a visit to Walley World.

Chapter 8

THE ONCOMING TRAIN

After the bright lights of Vegas and overstimulation of the Consumer Electronics Show, I was looking forward to a change of pace. So at the end of January 2016, I flew to Salt Lake City, Utah, and then drove east into the Wasatch Back region of the Rocky Mountains. I checked into a hotel in a small town, put on my snow boots, and started walking.

A few hours later I was watching the sun rise over a quiet mountain lake, admiring snowcapped peaks off in the distance, and listening to a gentle piano sonata playing on headphones. I felt relaxed and content. And then the tranquil scene was interrupted.

A long, low whistle, like breath exhaled through clenched teeth, echoed across the water. A plume of thick gray smoke traced its way along the distant shore, and then it turned, moving closer. It took me a moment to realize what I was looking at: an old steam locomotive, speeding across the lake's surface, straight toward where I stood. I watched, transfixed, as it approached, and then moments before reaching me, it exploded, transforming into a flock of black birds that swarmed around and past as the music came to a crescendo.

When I took off the headphones and Gear VR headset, I was still in the mountains but at a different setting—a demo room in the New Frontier Exhibition at the Sundance Film Festival in Park City, Utah. The New Frontier program started in 2006 as a showcase for cutting-edge media and technology at the annual gathering of indie filmmakers; ten years later, it was dominated by virtual reality experiences, and I was there to check out the artistic side of the VR revolution. Tranquility and relaxation would have to come later.

———

The exhibition was located in a gallery space on two floors of a hundred-year-old hotel on Park City's Main Street, and home to more than two dozen different stations outfitted with Rift, Vive, and Gear headsets. Many of the VR experiences on display amounted to simulated environments, like the mountain lake of Chris Milk's *Evolution of Verse*. Another demo, Jake Rowell and Ben Vance's *theBlu: Encounter,* took me on a dive to an underwater shipwreck, where I watched as fish and rays and ultimately a massive blue whale swam around my head.

These experiences had action and setting but no real narrative or plot. They reminded me less of modern films than of nineteenth-century magic lantern shows, or the very first moving pictures. The idea seemed to be to emphasize the immersive nature of VR rather than tell a particular story.

Other experiences followed more traditional narratives, like Janicza Bravo's *Hard World for Small Things*, a dramatic short that put the viewer in the backseat of a convertible driving around Los Angeles, watching a group of friends who get involved in an accidental shooting. It exploited the first-person presence of VR to create a kind of you-are-there thrill, making the shooting seem a little more real and visceral—but since the video was just from a static 360-degree camera and the scenes lacked any real interactivity, it felt less like a great leap into a new medium and more like a very fancy 3-D movie.

But as I passed from station to station trying out the different VR experiences, one of them did truly surprise me. Sylvain Chagué and Caecelia Charbonnier's *Real Virtuality: Immersive Explorations* was a multiuser immersive simulation built around a real-world stage set—a roped-off section of the room with some odd-looking boxes and props strewn around the space. As I watched from the line while waiting to try it out, two participants donned Rift headsets modified with multiple protruding antennas and white tracking dots, a pair of tracking gloves, weird devices strapped to their shoes, and large white backpacks filled with wireless computers. They looked more like astronauts on a spacewalk than attendees at a film festival and, as they went through the ten-minute demo, seemed to be engaged in some equally high-tech and alien activity. They traced odd paths around the floor, leaning and twisting around obstacles that only they could see, occasionally picking up one of the strange-looking props and carefully inserting it into a hole in a box resting on a pedestal in the middle of the room.

On my turn, I was paired up with another festival attendee, a middle-aged advertising executive named Matt, and we helped each other strap on the equipment.

When the headset powered up, I found myself on the bridge of a futuristic spaceship, stars shimmering out one window and Earth spinning by on the opposite side, but when I turned to where Matt had been standing, he had been replaced—instead, there was a small being with a shaved head, wire-frame glasses, and a skintight black space suit. I raised my hand and waved at it tentatively.

The avatar cocked its head and raised its own hand, waving back, and then I heard Matt's voice from across the room. "Hi. Wow, is that really you?"

Chagué and Charbonnier's Sundance demo was designed to offer a room-scale VR experience, something like the HTC Vive, but it used a custom setup of twelve infrared cameras tracking small sensors attached to the headset and to each player's feet and hands. Objects in the real world were also marked with the sensors, allowing them to appear in the virtual world and be manipulated and tracked too. In virtual reality, the box I'd seen on a pedestal floated in midair in the spacecraft, like some kind of futuristic flight control.

A narrator's voice invited us to explore the space, and I took a tentative step forward, like a baby learning to walk. Looking down at my feet, I could see two legs clad in a black space suit—not my legs, but they seemed to match my movement, so I inched forward, shaky at first, then took long and clunky Frankenstein stomps around the room.

As I got used to my new body, the narrator explained that the floating box was a time machine, and that we had to find special crystals in order to activate it. Clomping around, I spotted a glowing yellow mineral on the floor near the ship's pilot seat—an obvious illusion, but when I reached down to touch it, something was actually there. In the real world, it was just a piece of wood with tracking dots glued onto it, but the tactile feedback of actually touching something real and seeing it in the virtual was profound. In that moment, I bought into the illusion entirely: I was on that spaceship, and I was about to activate a time machine.

From behind me, I heard Matt shout that he'd also found a crystal, and then we both turned and made our unsure steps toward the box in the center of the room. Color-coded holes in the box glowed in anticipation, indicating where we should place the control rods, and we both inserted them, holding our breath. A whooshing sound filled the room, and everything turned white.

When the bright lights faded, I was standing somewhere else—on the deck of an open-air platform moving through the urban canyons of a futuristic cyberpunk city. Neon signs for strange products and unimaginably tall skyscrapers flashed past. Matt and I explored another scene, this one with two virtual characters dancing on the craft enjoying a date, before finding two more crystals and transporting ourselves to our final—and most convincing—virtual location.

We were in some kind of underground cavern, standing on a stone platform above a floor studded with glowing stalagmites. In front of us, the time machine had now opened to hold up a wooden torch, already ablaze. Matt's hand reached for it and then recoiled, like he'd been burnt.

"Whoa," he said. "It's really there!"

Even though I was a little afraid that I might burn myself, I reached out and grabbed the torch by the handle. It wasn't hot, but it was still a shock to feel the virtual object in my hand. With a simple bit of stagecraft—one of the *Real Virtuality* staff must have placed the object in the box during the whiteout time travel animation—the team had provided a sense of physical proof that we'd been transported, and gave us a totem that made the illusion feel even more real.

I lifted the torch (in reality, just another stick decorated with tracking dots, animated in the virtual world to look like a light source) and waved it around, and the light from its flames illuminated the cavern. We were standing at the entrance of some kind of underground structure, like a subterranean Petra or the entrance to a haunted temple in a fantasy role-playing game. The effect was intensified when the narrator gave us instructions that sounded like something a Dungeon Master might say: "Find the magic stone at the end of the path . . . but before that you will have to face your fears."

In the words once spoken by a nerdy character on the TV show *The X-Files*: I didn't spend all those years playing Dungeons & Dragons and not learn a little something about courage. I held out the torch in front of me to light the way, and stepped toward the structure's rough-hewn stone walls. At its opening, the floor abruptly fell away, replaced by a rickety-looking wooden bridge suspended high above the rocky cavern floor. Two steps into this virtual world and I'd already have to face one fear—I'd have to make a daring Indiana Jones–style scramble across the gap and risk plummeting to my death.

I held my breath and my skin went clammy as I took my first tiny steps. The bridge looked so unsafe, like it could fall apart at any second, but it held and took my weight. My heart pounded as I inched forward step by step, and my head swam with a sense of vertigo, like I might actually lose my balance and fall over the side. But soon I was close to the stone floor on the far end, and with a panicked leap, I landed on the more solid floor and sighed in relief that I'd made it alive.

Then I heard a panicked voice from behind me, in the dark. "Hey! Don't leave me!"

In my anxiety to make it into the temple, I'd forgotten about my teammate, Matt. Turning around, I could barely see him standing in near darkness at the edge of the bridge, his virtual form slightly hunched, one arm reaching toward me, imploring for help. I stepped close to my edge of the bridge and held out the torch above the gap, giving him a little more light.

"I'm so sorry!" I told him. "It isn't that bad. Just walk across. I'll wait for you here."

Matt nodded, but his feet didn't move. He crouched slightly, and his head looked down, staring through the slats of the rickety wooden bridge. Even though it was expressed through a computer-animated avatar, the body language was clear: he was frozen in fear, afraid to take a step, petrified that he would fall from the passage and onto the rocky floor below.

I shifted my torch into my left hand, took a step back onto the bridge, and extended my right hand out toward him. "It's just a few steps," I said. "Come across, I'll help you."

After a few moments, he took a tentative step, placing one foot on the bridge. Finding it solid, he then dashed the rest of the way across, grabbing my hand and then pulling himself off the bridge and onto the solid stone platform.

We stood there a short while, still holding on to each other, catching our breath. Then without further discussion, we walked forward together. The passageway twisted and turned, and with only the light from my torch to show our way, there was darkness ahead until I was able to reach around each corner. At one point, the light illuminated a spider the size of my hand scurrying up a wall, and I recoiled from it, bumping hard into Matt as I swept the torch at it in an attempt to burn it. Then the corridor ahead was filled with cobwebs, and as I passed through them, I winced and tried to brush them out of my hair. We jumped over another pit trap, hearts once again racing, and then crawled under a fallen stone pillar, squeezing our bodies close to the floor in order to fit through the passage.

Finally, we reached the end of our journey: a stone platform looking out into a cavern full of glowing crystals. As the narrator congratulated us on completing our journey, a basketball-sized orb of fire floated toward us. Matt extended his hands, and it drifted into his grasp. A whooshing sound filled the room, and everything turned white again.

"You can take your gear off," a voice said. I'd almost forgotten. I lifted the Rift from my eyes and had a moment of shock when I saw I was still in the New Frontier demo room, not an ancient dungeon, and was standing next to a middle-aged guy wearing funny tracking gloves and a white backpack, not a space-suited time traveler. We'd both gotten lost in the illusion, and all the time we thought we were traversing rickety bridges, tiptoeing through spider-infested corridors, or crawling under obstacles, we'd just been pantomiming our way around an empty room while a long line of indie film fans stood around, waiting for their turn.

"That was insane!" Matt said with a huge grin on his face. "Wow, that was insane. Incredible."

"It was!" I replied. "I knew that I was doing a VR demo, but in the moment, it was so real. I was really worried I was going to fall into that pit."

"The bridge got me. I actually had to stop and think about it. That was insane. Bravo! That was incredible."

———

Going into Sundance, I'd assumed that the future of VR movies would proceed something along the lines of the history of traditional filmmaking. I figured the first VR films would mostly capture and replicate the real world, attempting to convey the experience of being in a place you couldn't really visit. Later attempts would look and sound better, until the simulations were basically indistinguishable from reality. If a VR documentary film promised that "you are there," it would actually make it happen. A VR drama would work much the same way, yanking the audience out of their theater seats and onto a set where they could explore and watch the characters. I imagined being an invisible presence in the room when the Godfather promised to make "an offer he can't refuse," an unnoticed visitor to the *Star Wars* cantina, an anonymous rider on the runaway bus in *Speed*.

It hadn't occurred to me that the real power of VR cinema is to completely obliterate the line between audience and performer. At the simplest level, an audience member can be a passive participant, a friend in the car, a person other characters talk to instead of just an unacknowledged viewer. At a more interactive level, their observations can actually cause the action—an actor could wait to speak until the viewer looks at him, or a scene won't start to unfold until the viewer walks into the right room. And then there are VR films that could play out like *Real Virtuality*, where the viewer is actually the star of the film; exploring a set, interacting with other characters, and overcoming obstacles. It's the difference between inserting the viewer in the Temple of Doom alongside Indiana Jones, and putting the viewer in Indy's actual place. The future of filmmaking isn't just the most immersive 3-D movies ever—it's a blend of cinema, video games, immersive theater, and fantasy role-playing. In *The Matrix*, we all are Neo.

And with that thought, I pulled my parka back on and headed back out into the snowy streets of Park City.

———

"VR movies—this is our mantra, this is our word, and this is going to be a landmark year," the man said, raising a hand to shade his eyes from the stage lights and peer out at the crowd. "For the first time ever, the general public will be able to watch VR movies in their own homes. This is going to be a game changer."

Colum Slevin's title was Head of Experiences for Oculus VR, but on that night his job was master of ceremonies for a party at the St. Regis Deer Valley hotel. Oculus had enticed Sundance attendees with the promise of "cocktails, light bites, good company and great content," and a strange brew of Hollywood insiders, visual artists, and virtual reality nerds had shown up. After about an hour of awkward mingling, Slevin kicked off a few presentations with a short speech and a promise that "the bar's gonna stay open."

Oculus threw the party to promote its cinematic VR division, Oculus Story Studio. It had been founded in 2014, after Brendan

Iribe ran a tech demo of the Rift headset for a major Hollywood director. "He took [the headset] off at one point and said, 'Brendan, let's make a movie. How do we do this?'" Iribe said. "But I didn't have an answer for him. We really wouldn't know how to do that. So we said, let's go back and study this, and figure out how to use the Rift for cinema."

To do that, Oculus hired Saschka Unseld and Max Planck, two veteran directors from Pixar Animation Studios, and put them in charge of a ten-person team that included alumni from other Hollywood powerhouses including DreamWorks and Industrial Light & Magic. Their mission was to develop a computer-animated film that could show off the Rift's storytelling capabilities, and hopefully inspire other filmmakers to start working in virtual reality.

Story Studio's first effort, a film called *Lost*, depicted a first-person encounter with a friendly robot in a forest and had debuted a year earlier at the 2015 Sundance festival. I'd seen it earlier and I'd been impressed. The short was only about five minutes long, but the virtual reality environment allowed it to be studied according to the viewer's own interests and pace.

When I was immersed in the movie, I could stop and examine areas of particular interest, and the environment would respond. I remember looking at a swarm of fireflies, and then they flew over to check me out.

In 2016, Oculus had returned to Sundance to show off another animated film, *Dear Angelica,* which I'd watched a few hours before the Story Studio party. The movie told the story of a young woman writing a letter to her mother, and it was like nothing I'd ever seen before. As I stood in her bedroom watching her work, her memories sprung to life all around me as painted, animated works of art. It was touching, beautiful, and surprisingly intimate: I felt like an intruder in both her bedroom and her thoughts.

At the cocktail party, the film's director, Saschka Unseld, and art director, Wesley Allsbrook, took the stage and showed how they painted the animations entirely in VR using a program called Quill—Oculus's answer to the painting program I'd tried out on the Vive, back at GDC.

"In *Dear Angelica*, when you walk closer to a memory, it will unfold and wind around you exactly the way Wesley drew it," Unseld said. "It would be impossible to create this if she had drawn on a flat page . . . we needed to create this inside of VR for it to truly be different, to truly be something that's unique."

Unlike the other films at Sundance, Oculus hadn't brought *Dear Angelica* to Park City in order to sell it to a distributor. Rather, the company was looking to sell filmmakers on the idea of VR, to show off the tools and to demonstrate how the medium enabled new ways to tell stories. The elevator pitch was simple: totally immersive movies that the viewer can interact with and explore.

To help make the case to Hollywood, Oculus enlisted a celebrity guest to close out its party. Joseph Gordon-Levitt—the actor and director known for roles in films including *Looper*, *Inception*, and *The Dark Knight Rises*—took the stage after Unseld and addressed the crowd.

I'm not particularly experienced in VR, but I've had a chance here at Sundance to see some incredible work . . . and it's just so inspiring.

There's one particular storytelling device that I am personally quite interested in, being an actor, and that's the device of the protagonist.

You find a protagonist, of course, in all these different mediums . . . in the oral tradition, in the written word, in theater, or in movies. But protagonists function differently depending on technology. In a novel, the protagonist can go on and on for pages, just articulating their own internal thought process. You can't really do that in theater or in a movie.

So what's gonna be the protagonist's function in VR? From what I've gathered so far, it seems like there's a fork in the road. In some VR experiences, I am the viewer, I watch the protagonist. And then in other VR experiences, I *am* the protagonist. When you *are* the protagonist, that's completely different than watching the protagonist. I feel like that's where VR really kind of becomes its own self . . . it's doing something completely new.

Whenever a new technology emerges, usually the first stuff that people try are like the old ways. The first movies were sort of like moving photographs or plays. But as new generations came and inherited the technology, the art of moviemaking developed its own language and its own devices. And that took, like, a hundred years. A hundred years ago, there were people that were saying, "Movies? That's a novelty. That's not art."

The question now is, what are we gonna do in VR? I think the only way we're gonna figure it out is to just try stuff. That's why it's so cool just to see you guys making things and trying things, and seeing how it feels to experience them. It inspires the hell out of me.

After the speeches were over, I went to get a drink and ran into Gordon-Levitt on my way back from the bar. I complimented him for his recent appearance on the TV show *The Muppets* (he sang "Fly Me to the Moon" with Miss Piggy) and then remembered we were supposed to be talking about virtual reality filmmaking.

"In some of the demos I've done, I notice an intimate connection with the performance that you don't get in a movie theater," I told him.

He nodded in agreement. "I feel sometimes that it's more like theater than movies," he said. "There's such a diversity to VR, and that's the fun part. There aren't any conventions. If you go to a multiplex, all of the movies are basically following all the same rules, but I've watched so many VR experiences here at Sundance, and they all had different rules. It's such an exciting time."

I asked if there was a particular VR movie that really excited him.

"Chris Milk's *Evolution of Verse* . . . the one with the train," he said. "I knew I was standing there with a headset, but I absolutely physically felt scared that the train was gonna bowl me over. That's how people in movie theaters felt a hundred years ago, when the Lumière brothers put a camera right by the train tracks. Now you see people doing VR here at Sundance, and they're screaming and crying and falling over."

He laughed. "It's just so exciting. Most generations don't get to inherit this kind of open sandbox, where the rules aren't set, and it's time to play and figure it out."

———

The following morning, I shook off my open-bar headache and trudged through the snow to check out the other side of the virtual moviemaking business. The Finnish consumer electronics company Nokia Corporation had rented a ski lodge on the outskirts of Park City to show off a new piece of hardware they billed as the first-ever virtual reality camera made for professional filmmakers.

Until recently, recording any kind of video for immersive cinema had required new inventions and unconventional hardware. Early attempts at immersive filmmaking, like the Sensorama booth or Cinerama theaters, used unique wide-angle lenses and curved screens to draw in the viewer. Decades later, 360-degree videos (like the documentaries *The New York Times* created for its Google Cardboard project) were filmed with hacked-together rigs of eight or more separate digital cameras, each one pointed in a different direction and recording video that had to be stitched together on a computer, requiring massive amounts of slow and costly postproduction.

Nokia's Ozo camera was a simple—but costly—solution to that problem. The camera was a single orb of plastic, the size of a melon, studded at regular distances with eight embedded camera lenses. Each lens had a 195-degree point of view, so together they could see in every direction surrounding the camera. And since each lens was connected and run by the same software, the system could capture 360-degree video in real time, automatically stitch the images together, and then immediately record it to a computer's hard drive or stream it over a network. Eight microphones dotted around the exterior also captured stereo audio synced to the video capture.

When Nokia officially debuted the device two months earlier, they'd recorded a live band performing on top of the iconic Capitol Records Building in Hollywood, and streamed the video to attendees watching in VR at a party across town in Los Angeles. They'd also announced the list price at a staggering $60,000.

At Nokia's ski lodge, I took a few minutes to politely watch a VR recording of the concert, and then buttonholed Ramzi

Haidamus, at the time the president of Nokia Technologies, to ask him if there was really a market for such an expensive product.

"Yeah, absolutely," he insisted. "This is the only product on the market today that is purpose-built for high-quality VR capture in 3-D, audio, and video."

"But there are enough people who want to do that and who can afford to spend $60,000?"

"The first phone that we made cost $4,000, and back then that was a lot of money," said Haidamus. "You make your first products to learn, to get feedback, to see what works and what doesn't work . . . This is going to be the flagship, and then I imagine going into a second line of products that's more for the average consumer." Nokia, he argued, was already in talks with every major Hollywood movie studio, including 20th Century Fox, where the company's in-house futurist, Ted Schilowitz, had been a key partner in the development of the product.

Historically, Nokia was best known for developing mobile phones, I pointed out, and companies like Samsung and Google clearly saw the future of VR as happening first on mobile. So why didn't Nokia develop a cheap mobile VR viewer first, instead of a high-end camera?

"Initially, we did work on a viewer," he said. "We had an interesting experiment going on. But I felt that the market was too crowded, and it's gonna be so competitive. And I wasn't sure how much room there will be to differentiate, because some of the viewers are going to become standardized, just like televisions today." Every headset needs content, he said, but consumers don't need every headset.

As we chatted, I mentioned how I'd been at the Oculus Story

Studio party the night before, and how excited the young film-makers at the event seemed to be about making movies in virtual reality. "Do you see VR entertainment becoming a dominant form of media?" I asked him. "Are VR films going to put flat movies out of business?"

"I look at VR as a much broader opportunity," he said. "The biggest opportunities are going to be in commerce. How many times do house hunters see a photograph of a house they're interested in, but it turns out the photograph was of the best part of the house, and eighty percent of the house was left unseen until they go there in person? Imagine if the real estate agent had walked Ozo around there. You cannot hide anything from it. It sees everything in 3-D. It's gonna be huge in real estate. Imagine the opportunities in education, where kids are taken around the world on trips. You don't even have to leave your desk, you just sit in the comfort of the classroom, and you show them dangerous things. I see this as a huge market.

"But going back to your question about movies—I see VR as a complement to the current form, not a substitute. I love movies. I would love to see *Star Wars* made in VR, to visit another planet, to go to Mars. But people love to go sit in front of the big rectangular screen. This is gonna be a really interesting way to complement that."

———

Just a few hours later, at another Sundance cocktail party, the movie studio 20th Century Fox promoted a big-screen block-buster by offering their guests a virtual visit to another planet. The event was held in the VR Bar, a converted conference space in a Park City business center, and when I arrived it was dimly lit, filled with dance music, and crowded with fashionable people who weren't talking to one another. They sat on leather divans, wearing Gear VR headsets and bright red over-ear headphones, or perched in front of computers wearing Rift headsets and holding Touch controllers. Their heads cocked at strange angles, staring at nothing, observing a world inside the headset and not the party around them. Their bodies twisted and turned in response to virtual obstacles, sometimes bumping into real people, spilling their drinks, or jostling trays of canapés. Other attendees stood in line waiting for their turn, watching this strange silent dance with a mix of bemusement and incredulity.

Inside their headsets, the partygoers were taking a tour of Mars via *The Martian VR Experience*, a virtual reality short based on the sci-fi film *The Martian*. The interactive adventure allowed viewers to experience the perspective of the film's protagonist, astronaut Mark Watney, as he tried to survive alone on the surface of the red planet. The project was produced by the film's director, Ridley Scott, and was largely shot during the production of the movie, using the same actors and sets that would appear onscreen in theaters.

"It's a brand extension of the movie," said Mike Dunn, president of 20th Century Fox Home Entertainment. "It's really leveraging that content, going out during the theatrical window while you have a consumer's interest. They can get into [*The Martian*] either through VR or through theaters. I think that if you watch our VR experience, there's no way you won't go see the movie later, and if you see the movie and then find out there's a VR experience, you'll want to go to Mars and experience it for yourself."

The Martian VR Experience was only the latest VR film to emerge from the Fox Innovation Lab, a cutting-edge think tank run by the movie studio and dedicated to creating "true next generation home entertainment experiences." In 2016, the lab premiered a VR experience called *Wild*, tied into the Fox Searchlight film of the same name. The three-minute, 360-degree video starred Reese Witherspoon as her film character, Cheryl, in an original scene tied into the movie. I'd tried it a few months earlier, and was impressed when it drew me in using methods that would be impossible in a movie theater. In the short, Cheryl hikes up a trail through a sun-dappled forest, toward the stationary viewer, and when she gets close, she sits down on a large rock to catch her breath. The immersive video made the scene unusually intimate, and seeing Cheryl in 3-D in front of me made her feel like a real person. But *Wild* got real when Cheryl turned and looked directly at me; we shared a connection for a minute, until I realized she wasn't looking at me but *through* me.

I turned around, and discovered another character had appeared in the forest: the ghost of Cheryl's mother, played by the actor Laura Dern. Since the VR application knew which direction my head was facing, it was able to insert her into the scene when I wasn't looking, and wait until I was actually facing her to have her start talking. When she finished, she sat there until I turned back to Cheryl, and then she disappeared again while I wasn't looking.

It was a simple trick, but one that illustrated an exciting aspect of virtual reality: the idea that an experience can be passively interactive. Unlike video games, which require users to actively navigate through a conflict or narrative, VR films can allow the viewer to relax and watch but still control the action. It opens up interesting new ways to tell a story—like dialogue that changes depending on which character you're looking at, or alternate scenes that you can only see if you move your virtual self into a different room.

Most of the studio-produced VR to date had been immersive but static, essentially just 360-degree videos with no interactive elements. But increasingly, experiments like *Wild* and *The Martian VR Experience* included segments where users could take control. At the Sundance party, guests using a Rift headset and Touch controllers could drive Watney's rover across the Martian landscape, for instance. It wasn't the most audacious use of virtual reality, but the audience seemed to enjoy it. "This is so freaking dope!" a young woman wearing a headset and a backless knit dress told me right before she unknowingly grabbed my thigh, unable to see that I wasn't her boyfriend.

Interactive movies might work better at home than at a crowded party—and the movie studios were actually counting on consumers wanting to avoid crowds and stay on their own couch. Around the same time it debuted the *Martian* experience, 20th Century Fox announced a partnership with Oculus VR to release more than a hundred titles from its library to a virtual reality cinema called Oculus Video. The idea was to offer viewers a big-screen experience without leaving their homes; by putting on a Rift or Gear headset, users could sit in a simulated theater and watch films like *Die Hard*, *Office Space*, and *Alien*.

While much of Hollywood planned for VR to become the next big thing in filmed and scripted entertainment, other entrepreneurs bet that VR would upend the way people watch live events.

"Several years ago I began tracking VR, and I recognized this is not just a parlor trick," said Peter Guber, chairman and CEO of Mandalay Entertainment and co-owner of four professional sports teams, including the NBA's Golden State Warriors and the MLB's Los Angeles Dodgers. "It's a new technology to connect audiences and artists so that people who can't make it to the arena can still feel like they're in a front-row seat."

I'd spoken to Guber six months before Sundance, after he joined the advisory board for virtual reality start-up NextVR, which was building a platform to allow users of any VR headset to tune into live events and watch the action through 360-degree cameras located onstage or on the sidelines. "We did a test where we [had VR cameras] at an event in the San Francisco Bay Area at a Warriors game," Guber said, "and when we streamed it back to New York so that we could see how it worked and how effective it was, it was so beautiful. VR puts you in the front-row seat. On the glass in a hockey game, or on the red carpet at an awards show, or in the front row of a basketball court . . . you're sitting there virtually, and looking at what you want. It's the closest thing to physically being in the stadium."

NextVR had already streamed a handful of live virtual reality events, including basketball games, soccer matches, and live concerts by bands like Coldplay, but was competing with similar VR start-ups and old-school media giants to lock down VR content agreements with sports leagues and record labels. "There's deals to be made and legal requirements that have to be met, but VR isn't coming, it's already here," Guber said.

I'd been skeptical about that statement when I talked to Guber, but the scene at Sundance convinced me he was right. I could already watch a 3-D movie built for VR, enjoy a 2-D classic in a virtual theater, or experience a live sporting event from the side of the field—all without ever leaving my couch. And when I looked around the VR Bar, I didn't see people playing with a new toy; I saw people embracing a new medium, getting lost in it, and doing it as naturally as they'd turn on a TV.

And I saw them leaning forward. Film and television are lean-back mediums, passive experiences you consume as you relax and sink into a chair or a couch. But the people consuming VR were alert and active, feeling their way around this new world, active participants in their stories. They were fully engaged and completely immersed.

I suspect that television and film aren't going to go away. New media doesn't always replace what came before it—we still listen to radio and we still go to the theater, after all. But the experience of VR is so compelling that it's hard to imagine it won't become our go-to medium. Right now it's an emerging art form, a new frontier, a train off in the distance . . . but it's coming closer. And when it arrives, we're going to look at flat-screen media in the same way we look at black-and-white silent movies.

Chapter 9

THIS IS REAL

It was a cold and cloudy day in San Francisco, but Palmer Luckey was dressed for the beach, as always, wearing a Hawaiian shirt, sandals, and cargo shorts. The few dozen reporters crowded around him wore jackets and hooded sweatshirts, even though we were indoors.

"Hi, everybody! I'm Palmer, the founder of Oculus," he said, as if we didn't know. "Thank you all for being here. I think we're showing forty-one games here today, thirty of which are going to be launching on March twenty-eighth with the Rift."

It was March 14, 2016, and the final consumer version of the Oculus Rift was due to be released in just two weeks. Oculus had organized this demo day to show off the hardware to members of the media, allow them to meet some developers who were making virtual reality experiences, and give everyone a chance to experience the launch version of the Rift for themselves.

"I'm glad that you guys all came," Luckey continued. "It's pretty cool going from a few years ago, when I was lucky to get any press to talk to me and we were out there trying to convince developers this was the future, to now when we have tons of development in VR and tons of journalists here trying it out."

The day was set up so that each reporter had up to a dozen short sessions scheduled, each one with a different VR developer and a demo of their soon-to-be-released VR game. I started off with Chronos, a fantasy role-playing adventure rendered in third-person 3-D, making me an invisible observer watching my animated character hack and slash his way through a monster-filled dungeon crawl. After that, I moved on to a first-person experience, a futuristic shoot-'em-up called Damaged Core, where I saw through the hero's eyes as I blasted alien robots with twin laser guns. And then I tried The Climb, a first-person rock-climbing game that simulates the scramble up a series of sheer cliff walls.

I was doomed from the very first rock. The Climb wasn't unusually realistic—the computer-generated mountain environments looked handsome but artificial, and when I gazed where my body should be, I saw only a pair of cartoonish disembodied hands. But after I pressed a few buttons on the controller to grasp a crack in the cliff face and start to pull myself up, I made the mistake of looking down. Even though it was obvious that I wasn't actually hanging on to the side of a mountain, some primitive part of my brain concluded I was actually a few thousand feet up. My stomach lurched, my skin went clammy, and I immediately felt like I was going to throw up.

I pulled the Rift headset off and squeaked out a half thank-you, half apology to the developer, and then I lurched over to a nearby lounge area and collapsed onto an overstuffed couch. Waves of nausea washed over me as I lay there, eyes closed, taking deep breaths, waiting for the vertigo to wear off.

After a few minutes, I felt stable enough to open my eyes again, and when I did, I noticed Luckey sprawled on the couch opposite me, feet up on the cushions, watching me come back to my senses.

"What got you?" he asked.

"The Climb."

"Oh man, that's crazy good," he said. "What are you, afraid of heights?"

"No."

"I see," he said, and grinned.

A few weeks later, Palmer Luckey took his own trip to the mountains. On March 26, two days before the consumer version of the Rift hit stores, Luckey traveled to Anchorage, Alaska, to personally deliver a headset to Ross Martin, a VR enthusiast who was the first customer to preorder the hardware when it went on sale three months earlier.

"I got an e-mail that was very nondescript and low-key, and it just said that 'hey, we want to ship your Rift to you on Saturday, we want to deliver it to you, please call us to confirm some details,'" Martin told the gaming news website Polygon. "I had no idea Palmer would show up."

When he did, Luckey live-streamed the entire visit to Facebook. In the archived video, Luckey has a slightly glazed, jet-lagged look about him, but he wears a big grin—and his customary Hawaiian shirt and cargo shorts, of course. He chats with Martin and watches him unbox his Rift headset, wearing an undisguised expression of pride.

"This is incredible," Luckey says. "I've been working on this thing for so long, and you're the first person to actually get one, so it's kind of like me taking all that hard work, and handing it off to you, so you have to make sure to have fun with it."

After less than five minutes of small talk, Luckey made a quick exit. "Now I need to get home as fast as I can so that I can keep working on the launch," he said.

It was a comically short visit for about twenty cumulative hours of travel time, but Luckey said he felt it was something he had to do. "I was pretty adamant," he told Polygon. "I said, 'Hey, guys, I've been working on Oculus since 2012. I'll be damned if some random delivery guy is going to get the satisfaction of delivering the first Rift. That's mine.' I figured I could take a day out of the launch process for a quick vacation."

On his way home, Luckey posted a travel update on his Twitter account. "Hiking through Alaska in the winter wearing flip-flops was a mistake," it said.

———

Meanwhile, in New York City, it was late at night and my dog wanted to go for a walk, but she had to wait, because I was in outer space. My fighter squadron was on an escort mission through enemy territory, and I had to stay alert. Peering out the window to my left, I saw one of the capital ships we'd been assigned to protect—a massive vessel that filled my entire field of vision. Turning to the right, I saw the rest of the fleet and, beyond those ships, countless stars. I was admiring the view when the radio blared with the excited voices of allied pilots. I looked up, and an enemy fighter overtook me from behind, its cannons silhouetted against the stars.

It was easy to lose yourself in the immersive fantasy created by the Oculus Rift, which began shipping on March 28, 2016, and started arriving at homes and offices in over twenty countries a few days later. It was only when a zero-gravity dogfight was interrupted by the poking nose of an anxious dog that I realized virtual reality was finally real.

At first glance, the first consumer version of the Oculus Rift didn't look much different from the VR hardware I'd seen before: a pair of goggles with a molded plastic head strap and built-in headphones. I plugged it into my computer, pulled it down over my eyes, and peered through the lenses at a 1080-pixel high-resolution AMOLED screen. But unlike previous headsets, this Rift looked and felt like high-end consumer electronics, not a science experiment gone wrong. It was covered with a fine woven mesh that made it feel soft and inviting; it was lightweight while still seeming substantial; and it adjusted to sit on my head without pressing uncomfortably into my face. The external tracking camera that plugged into the computer and watched my head movements was stylish too. The whole system felt like something I could be proud to display in my living room, not want to hide under a desk.

The setup process was equally well designed. The Rift headset shipped in a cleverly designed box with the tracker, a custom remote control, and a Microsoft Xbox game controller. New users were directed to open up a link in a web browser to download the required software, and it walked them through installation, fitting, and customization of the system. Even though I was certain that almost all the initial users of the Rift were VR enthusiasts and gadget fiends, the software felt like it was designed to be accessible to anyone. After watching a short VR movie, users were introduced to an equally easy-to-navigate control interface, presented as a virtual living room called Oculus Home.

From Home, users could launch into a variety of virtual worlds. The Rift launched with more than thirty games available in its store, and they ran the gamut from high-energy space combat to platform jumpers, puzzles, and pinball. Crucially, Oculus Home allowed users to sort these games by comfort level, so if someone was new to VR, they could choose a "comfortable" experience with little movement and a stationary camera, and not be overwhelmed by an "intense" game—like The Climb—that might make them lose their lunch.

When I told people that I had ordered a Rift, invariably the first question they asked was something along the lines of "Doesn't that make you sick?" But the reality was that the hardware design did away with many of the problems that cause motion simulator sickness. For one thing, the headset had such a wide field of view that when I wore it, there was no obvious end to the display; it convinced my brain that I was present in the virtual environment, not looking at a screen. The Rift's positional tracking system also worked seamlessly. When I turned my head, it followed the movement and smoothly displayed new parts of the environment, without rendering problems or lag. The image looked normal, so the experience felt normal.

I am highly susceptible to motion sickness, both in the virtual world and the real one. I can't go on motion-simulator rides at amusement parks. Watching a movie with excessive shaky-camera movements will sometimes make me ill. I can't even play first-person video games on a TV in my living room without eventually getting queasy. But when I stuck to "comfortable" experiences with the Oculus Rift, I never had any kind of nausea. My brain believed it was real, so it wasn't a problem.

Maybe that's why some of my favorite experiences at launch were a little old-school. The game I became addicted to first was Hidden Path Entertainment's Defense Grid 2: Enhanced VR Edition, a tower defense game—a real-time strategy genre that's most popular on mobile devices but which worked brilliantly in VR. While wearing the Rift headset, I could see the game as if it were playing out on a miniature battlefield floating in space in front of me. I felt like I was playing a tabletop war game, but instead of pushing around inanimate tin soldiers, there were fully animated units that actually moved and fired their guns.

Another standout launch title was Lucky's Tale, a game developed by Playful Corp and published by Oculus VR, which came bundled for free with every Rift. It was a charming cartoon-style platform adventure, and it played a lot like a Mario game would if you could actually enter the Mushroom Kingdom—the playable environment was all around you, and the heroic cartoon fox navigated hazards that appeared to have real depth and weight.

And then there were more action-oriented VR experiences, which, even though I was still a bit too prone to motion sickness to fully appreciate, were just too cool to avoid. CCP Games' EVE: Valkyrie was the space combat game that had my dog begging for relief—it came bundled with every preordered Rift, and it was ridiculously fun, with gorgeous graphics and frenetic action. It was also rated "intense," and I found I could play it for only a few minutes before feeling queasy, but it was so exciting I kept returning and trying to build up my tolerance.

These amazing experiences didn't come cheap, and that was the biggest strike against the Rift. The headset and accompanying peripherals retailed for $599, and they required a pretty fast computer to work. Customers could expect to lay out at least $1,500 for a new PC plus the headset and software, and as much as $3,000 for something closer to top-of-the-line.

And for all that money, the Rift experience wasn't perfect. The headset fit comfortably on my head, but the shape of it left a gap near my nose where outside light could bleed in. And while motion sickness could be controlled and avoided, eyestrain remained an issue—I had to take the headset off every half hour or so to give my eyes a rest. And even though there were some very fun games available to play at launch, the platform was still new, so the Rift's software catalog was thin at best.

The Rift also suffered from the lack of motion-tracked, built-for-VR controllers. At launch, the system and all its games had to be navigated with an included remote control or Xbox control pad. They worked, but VR begs for controllers that replace your hands. More often than not, the first thing people did when they put on a Rift was hold their hands out in front of the headset, and it was a letdown when they saw nothing there. Oculus had already solved this problem with their Touch controllers, but they weren't set for a commercial release until the second half of 2016. Without them, the Rift felt incomplete.

Some critics dinged Oculus because the Rift didn't support room-scale VR, the main selling point for the HTC Vive. But at launch, I didn't miss that on the Rift—partly because I didn't want to get out of my chair, but mostly because the majority of VR experiences didn't really need room-scale support. I was used to consuming the majority of my entertainment while basically stationary—on a couch or in my computer chair—and I didn't see that changing anytime soon.

But even though it couldn't do everything, the Rift was shockingly complete and polished, especially for a first-generation product. Oculus managed to produce a device that could thrill gamers and VR enthusiasts, but was also ready for the mainstream. While it was so new and expensive I couldn't recommend it to the average consumer, I still couldn't wait to show it off to everyone who came to my house, old or young, geek or luddite. And I suspected in time—as Oculus refined the product, and as more game designers, filmmakers, and artists created content designed for VR—a future version of the hardware would break into living rooms everywhere. I didn't know if that would happen in a year or a decade. But I knew when it happened we'd remember the Rift as the device that kicked off the age of VR.

———

On April 5, just eight days after the Rift was released, the HTC Vive hit stores. Compared to the Rift's high-end feel and elegant design, the Vive was an example of utilitarian geek chic. The headset resembled a prop from a 1950s alien invasion movie, and the wireless controllers looked like lightsabers that belonged to a Sith Lord. The setup process wasn't particularly streamlined, either. The Vive arrived in a case nearly twice the size of the Rift and with far more components inside—the headset, the controllers, two base station sensors, and a link box to plug all that into before connecting to my computer. There were cables everywhere and, alarmingly, five different power cords—I had to unplug half of the electronics in my living room in order to free up enough outlets just to get the Vive powered on. And because the Vive was designed for room-scale VR, its two sensors had to be positioned close to the ceiling, in opposite corners, in order to get an overlapping, unobstructed view of the room. Fortunately, the system came with wall mounts and very long cables, but setting up the sensors took me another hour of pulling cables behind furniture and drilling holes in my walls.

Finally, once the system was powered up and operational, the Vive's setup software required that I repeatedly walk around the room with one of the controllers, tracing out the borders of my "play area," or the space where the Vive would place its out-of-bounds-indicating Chaperone walls. Once again, my dog became a victim of the age of VR; I had no choice but to drag her bed into the hallway in order to give myself room to move around. At least with the headset on, I wouldn't be able to see her aggrieved hound-eyes.

All the work was worth it once I actually got the Vive running and began to explore a collection of VR games developed by Valve Corporation called The Lab. Designed as a showcase for the system's capabilities, the software presented a series of mini-games set up around a laboratory setting from Valve's beloved and popular Portal franchise. In one game, I played the part of an archer standing on the battlements of a castle under siege from a rampaging horde; the controller in my left hand appeared in VR as a simple wooden bow, and my right hand transformed into a red feather-fletched arrow. When I moved my hands in the real world, the weapons did the same in the game, and I instinctively knew what to do: pantomime the motions of nocking an arrow, pull it back, and let my shot go. The thrill when I hit my target was like an electric shock. Even though I'd done it before, being able to manipulate objects made VR vastly more immersive, interesting, and fun.

After exploring The Lab for a while—and repeatedly untwisting myself from the long, heavy cables that trailed from the back of the Vive headset to my PC—I fired up one of the system's launch titles, a puzzle game called Fantastic Contraption. Originally developed in 2008 as a web-based browser game, Fantastic Contraption required players to build simple machines using a tool kit of items like rods, wheels, and motors with the objective of getting the machine to roll a few feet to a goal. As a two-dimensional web game, it was intriguing but forgettable; reinvented for an immersive virtual environment, it had me immediately hooked.

In the simulation, I saw a pastel-colored cartoon-style world, something like a miniature golf course floating in a blue sky, where I was meant to build machines to traverse the unique terrain of each puzzle. I couldn't see my own body when I looked down at myself, but the two Vive controllers were floating in space, and a brief audio tutorial showed me how to use them to build a contraption: grabbing parts from a toolbox, pulling and stretching those objects to the preferred size and shape, and then snapping them together to form axles, frames, and entire vehicles.

Like the archery game in The Lab, Fantastic Contraption initially drew me in with its control system; it felt like I was actually picking up and putting together physical objects, not just waving my hands around in the air. But the game really won me over when I remembered I could actually move around in the virtual space. I could put my contraption down on the ground, walk

197

around it, examine different sides, even walk back and forth on the course itself, checking out the obstacles between the starting line and finish point.

———

The room-scale experience was impressive, and I loved having controllers that actually tracked my hands. But in the weeks after the two big VR headsets launched, I found myself using the Vive less and less. It was a pain moving away furniture (and herding away curious dogs) every time I wanted to roam around in one of the Vive's virtual worlds. If I had a room in my house totally dedicated to VR, maybe that wouldn't be such a big deal. But even in that case, I suspected that the heavy, dragging cables would continue to be an annoying problem.

Instead, I returned to the comparatively limited experience of the Oculus Rift. Aside from my continuing addiction to the tabletop appeal of Defense Grid 2, I found that I'd reach for the Rift for short bursts of VR as a diversion, or after I read about something new that might be fun. The headset was light and comfortable and sat on a shelf right next to my desk. If I saw a news story about, for instance, a new VR film that had been released, I could reach over, pop the headset on my head, find the video in Oculus Home, and finish the entire experience in just a few minutes without ever getting up from my chair. The Rift experience was easy, polished, and comfortable. It was a home entertainment device, while the Vive still felt like an exciting but complicated piece of laboratory tech.

Still, it was clear that the Rift owed a lot to the Vive. There had been a huge jump in quality between the Rift's first and second development kits—a period that coincided with Valve's attempts to work with Oculus, and the eventual move of Michael Abrash and Atman Binstock from Seattle to Silicon Valley. For instance, early versions of the Rift relied on internal sensors for tracking the motion of a headset, while Valve was already using computer vision and external cameras.

A few months after the consumer versions of both headsets were released, Alan Yates, one of the Valve engineers responsible for developing the company's tracking system, argued the point in a post on Reddit. "Every core feature of both the Rift and Vive HMDs are directly derived from Valve's research program," he wrote. "Oculus has their own CV-based tracking implementation and [lens] design but the CV1 is otherwise a direct copy of the architecture of the 1080p Steam Sight prototype Valve lent Oculus when we installed a copy of the 'Valve Room' at their headquarters. I would call Oculus the first SteamVR licensee, but history will likely record a somewhat different term for it."

———

Any controversy over who invented what did little to dis-
courage investors and entrepreneurs from trying to make their own mark in VR. According to an analysis by market research firm Digi-Capital, venture capitalists invested almost $1.2 billion in virtual and augmented reality businesses in the first quarter of 2016 alone.

Meanwhile, VR began to appear in pop culture to a degree that hadn't been seen since the early 1990s. In April, television's top-rated show, *The Big Bang Theory*, began an episode with main character Sheldon Cooper taking a virtual walk through a forest using a Zeiss VR One headset.

After years of caution and skepticism, some of the biggest companies in the entertainment business had finally started to get serious about making content for VR. Disney Studios released Disney Movies VR, an app for both the Rift and Vive headsets that let users visit simulated environments from its movies and theme parks, as well as watch 3-D clips from movies including *The Jungle Book*, *Captain America*, and *Star Wars*. The broadcast network NBC announced it was teaming up with Samsung to simulcast the 2016 Rio Olympic Games in VR. And the cultural giant *The Simpsons* celebrated its six hundredth episode by turning its regular show-opening "couch gag" into an immersive 3-D *Planet of the Apes* parody for Google Cardboard, called "Planet of the Couches."

VR was winning awards too. In May, the Cirque du Soleil virtual reality experience I'd enjoyed on my way to Las Vegas won an Emmy Award for its creators, Cirque du Soleil Média and Felix & Paul Studios. And in June, the *New York Times* VR documentary *The Displaced* won a Grand Prix at the Cannes Lions film festival.

It seemed like everyone was trying VR. In October, in an appearance at Oculus's annual Connect developer conference, Mark Zuckerberg boasted to the crowd that when world leaders stopped by Facebook's headquarters to meet with him and find out more about the massively influential company, he'd jump at the chance to show them virtual reality for the first time.

"Depending on what kind of a leader it is and what their culture is, maybe we'll play a first-person shooter," Zuckerberg said. "By the time the headset is done and we take the headset off, they're just amazed, and they don't want to leave.

"I actually had this one situation where the wife of a prime minister was yelling at her husband that he had to leave, go catch the plane home, and he was just sitting there saying, 'I was told there was a dinosaur. I demand to see the dinosaur,'" Zuckerberg said.

Even the classic pen-and-paper fantasy role-playing game Dungeons & Dragons made the leap off the tabletop into the virtual world. On November 16, AltspaceVR announced a partnership with D&D publisher Wizards of the Coast (a subsidiary of Hasbro) to bring officially licensed D&D assets to a tavern-themed VR game room. Users could gather around a 3-D table and control individual avatars representing each player. They could build a map using dungeon-, wilderness-, and city-themed terrain tiles; record their characters on official D&D character sheets; and move around figurines representing various D&D player classes, as well as monsters like dragons and gelatinous cubes.

———

The market heated up even more on October 13, 2016, when Sony shipped the results of its Project Morpheus—a $399 headset called PlayStation VR. Priced in a sweet spot between the high-end Vive and Rift and a plethora of low-end phone-based viewers, the PSVR was Sony's bid to take over the entire middle of the market by sneaking into homes through the living room.

To be fair, the device wasn't inexpensive—it sold as a peripheral to the $399 PlayStation 4 video game console and required a $59 PlayStation camera and one or two $25 PlayStation Move controllers. But Sony wasn't marketing the device to new users who needed to buy all that from scratch. At the time, the company had sold almost 50 million PlayStation 4 consoles worldwide, and it was counting on those video game enthusiasts to embrace the PSVR as a mainstream competitor in the virtual reality race.

The design of the PSVR reflected Sony's ambitions to make it that year's cool must-have toy. The headset was made of white and black plastic, and the exterior was dotted with a constellation of LEDs that glowed blue when the headset was in use—a striking futuristic look that made it stand out in a living room and made anyone who wore it look like an extra from *Tron*. The display was attached to a hard plastic shell that fit over the user's head like a construction helmet, instead of using ski mask–style straps like competing products. This change made the headset slightly less immersive, since it wasn't held flush against your face, but a lot more comfortable to wear for long periods of playing video games.

The headset also used less-expensive components. Instead of two screens producing a resolution of 1080 by 1200 pixels for each eye like the Rift and the Vive, the PSVR had one screen with a resolution of 1080 by 960. As a result, PlayStationVR experiences were less vivid and realistic than what competitors could offer. And while the video game system's Move controllers offered an option for hand tracking, the flashlight-sized wands weren't as versatile or intuitive as the Vive controllers or Oculus Touch.

All considered, the PSVR offered an experience best categorized as "good but not great"—comfortable and fun, but definitely less impressive than what you might experience using one of its PC cousins. For the money, though, it was an excellent option, and a reasonable holiday wish list item, at least for gamers within the existing PlayStation ecosystem, or with very generous loved ones.

Expectations were high going into the PSVR's launch.
Earlier in the year, market research firm SuperData Research estimated Sony would sell 2.6 million units between its October debut
and the end of 2016—a massive number that was likely to dwarf
even the combined sales of the high-end competitors. While neither HTC nor Oculus would say how many headsets they'd sold so
far, the best estimate available indicated that the two companies
had moved just 300,000 units combined in the six months since
their launches, and would likely reach only about 775,000 by the
end of the year. That's a respectable figure for very expensive computer hardware in a brand-new market, but hardly a blip on the
radar of the top products in the consumer electronics market. In
comparison, at the time, Apple was estimated to sell more than
667,000 iPhones every single day. (To be fair, the huge popularity
of smartphones did help drive much stronger sales of entry-level
phone-based VR viewers; the same report estimated that Samsung
would sell 2.3 million $99 Gear VR viewers by the end of 2016.) If
the PSVR managed to sell anything close to the millions of units
that SuperData predicted, it would give Sony a huge lead and go a
long way toward making VR a mainstream technology.

Meanwhile, Oculus was still the best-known and most-buzzed-
about brand in the market, but cracks were starting to show in the
organization, and its reputation was beginning to tarnish. In September, news website *The Daily Beast* revealed that during the
final months of the hotly contested 2016 US presidential elections,
self-described libertarian Palmer Luckey had funded a pro–Donald
Trump political organization called Nimble America that was

dedicated to "shitposting" on the Internet and spreading anti–
Hillary Clinton memes—including a billboard outside of Pitts-
burgh that featured a portrait of the Democratic Party candidate
alongside the words *Too Big to Jail*.

The news came to light after a member of the group made a
post announcing their existence to a pro-Trump discussion group
on Reddit ("We conquered Reddit and drive narrative on social
media, conquered the [mainstream media], now it's time to get
our most delicious memes in front of Americans whether they
like it or not," it teased). Then a few hours later, a user named
NimbleRichMan made a follow-up post promising to match dona-
tions to the nonprofit group for the next forty-eight hours.

"You and I are the same. We know Hillary Clinton is corrupt,
a warmonger, a freedom-stripper. Not the good kind you see
dancing in bikinis on Independence Day, the bad kind that strips
freedom from citizens and grants it to donors," NimbleRichMan
wrote. "I reached out to the leaders of this community because I
am doing everything I can to help make America great again. I
have already donated significant funds to Nimble America, and
will continue to do so . . . am I bragging? Will people be offended?
Yes, but those people already hate Donald. They cannot stand to
see successful people who are proud of their success."

The posts struck some readers as a scam, an example of an anonymous group trying to capitalize on growing alt-right sentiment to
make a quick dollar. So when *The Daily Beast* contacted Luckey to ask
if he was NimbleRichMan, he confirmed to the reporters that the
group was real and he was its main financial backer.

"I've got plenty of money," Luckey told the site. "I thought it
sounded like a real jolly good time."

After *The Daily Beast* broke the story, it quickly became international news—the idea that a twenty-four-year-old tech titan worth $700 million was attempting to sway the presidential race using Internet trolls made Trump fans gleeful and most everyone else disgusted. Thousands of people posted on Facebook and Twitter that Luckey's behavior was turning them off from ever buying a Rift, and a handful of video game developers promised to stop making new games for the platform.

The next day, Luckey posted a statement on his Facebook page apologizing for the stunt, but denying that he was NimbleRichMan at all:

> I am deeply sorry that my actions are negatively impacting the perception of Oculus and its partners. The recent news stories about me do not accurately represent my views.
>
> Here's more background: I contributed $10,000 to Nimble America because I thought the organization had fresh ideas on how to communicate with young voters through the use of several billboards. I am a libertarian who has publicly supported Ron Paul and Gary Johnson in the past, and I plan on voting for Gary in this election as well.
>
> I am committed to the principles of fair play and equal treatment. I did not write the "NimbleRichMan" posts, nor did I delete the account. Reports that I am a founder or employee of Nimble America are false. I don't have any plans to donate beyond what I have already given to Nimble America.
>
> Still, my actions were my own and do not represent Oculus. I'm sorry for the impact my actions are having on the community.

In the hours after the statement went up, a few of Luckey's colleagues posted their own messages of support—or at least attempts at damage control. Oculus head of content Jason Rubin said he took Luckey at his word and "would not work in a place that I thought condoned, or spread hate." CEO Brendan Iribe promised that "everyone at Oculus is free to support the issues or causes that matter to them, whether or not we agree with those views," and reminded the public that "it is important to remember that Palmer acted independently in a personal capacity, and was in no way representing the company."

None of it did much to calm down the furor. In the days that followed, both news outlets and tech websites ferociously debated whether Luckey had done anything wrong, and even if he had, whether it was any of the public's business. A few prominent critics—including author and tech evangelist Robert Scoble—argued that Mark Zuckerberg should fire Luckey.

No such thing happened, but as Oculus entered the critical holiday gift-buying season, its formerly omnipresent, darling-of-the-press founder was suddenly conspicuously absent from corporate messaging and events. He didn't even show up at the third annual Oculus Connect convention in October, a decision that Oculus VP of product, Nate Mitchell, said he'd made on his own. "Palmer decided not to attend OC3 because he didn't want to be a distraction," Mitchell said.

The Nimble America story faded away in the overheated news cycle of the weeks leading up to the presidential election. But inside Oculus' owner, Facebook, executives began to tighten their control on the once largely independent subsidiary.

The first sign that things were changing came from Brendan

Iribe, who announced in a blog post on December 13 that he was stepping down as CEO to lead the company's PC VR group. "As we've grown, I really missed the deep, day-to-day involvement in building a brand-new product on the leading edge of technology," he said. "You do your best work when you love what you're working on. If that's not the case, you need to make a change. With this new role, I can dive back into engineering and product development. That's what gets me up every day, inspired to run to work."

Facebook didn't immediately announce a new CEO, and there was no real reason to believe that Iribe was sidelined because of the Nimble America fiasco. Instead, the change seemed motivated by a desire to restructure the company in order to run separate mobile and PC-based product groups. (On January 26, 2016, Zuckerberg appointed Hugo Barra, a former VP at Google and the Chinese tech company Xiaomi, to run the Oculus division as Facebook's vice president of virtual reality.) Still, the shake-up didn't help improve confidence in the company during the holiday season.

As 2016 came to a close, it became clear that the "Year of VR" had given the technology a somewhat lackluster start. Updated sales estimates from research firm SuperData indicated that the total VR industry shipped 6.3 million devices and earned $1.8 billion in revenue for the year. Low-cost mobile viewers accounted for the bulk of the sales, with the Samsung Gear VR moving 4.5 million units. The midrange PlayStation VR shipped an estimated 750,000 units, less than a third of what SuperData had estimated six months earlier. As expected, the expensive high-end headsets remained largely a niche product, but one surprise was that the Vive was ahead in the market: SuperData estimated that HTC had moved 420,000 units, compared with Oculus at just 240,000. Research firm

Canalys was slightly more optimistic, estimating 500,000 Vives sold to 400,000 Rifts, but still, the leader was obvious.

Life at Oculus didn't get rosier as 2017 rolled around. On January 9, the US District Court for the Northern District of Texas began a jury trial in the case of *ZeniMax Media v. Oculus*, the lawsuit alleging that John Carmack had illegally misappropriated trade secrets when he jumped from ZeniMax to Oculus, that Luckey had violated nondisclosure agreements, and that Oculus had exploited ZeniMax intellectual property and then refused to compensate the company.

The trial lasted eighteen days and featured all the major players in the Oculus saga. Carmack was the first to testify. While he admitted to taking code he had developed from one company to the other, he vehemently denied that he had used it in any way during the development of the Rift or other Oculus products.

On January 17, Mark Zuckerberg took the stand and echoed Carmack's testimony. "We are highly confident that Oculus products are built on Oculus technology. The idea that Oculus products are based on someone else's technology is just wrong," Zuckerberg said.

The Facebook CEO's testimony also revealed some new information about the acquisition of Oculus VR. Initially, Zuckerberg said, Luckey and Iribe wanted $4 billion for the company, but the Facebook team was able to negotiate them down to a $2 billion price tag plus $700 million "in compensation to retain important Oculus team members" and an additional $300 million in incentives—meaning that if Oculus managed to hit those unknown milestones, the final purchase price of the company would have been $3 billion, not the $2 billion initially reported.

Zuckerberg also testified regarding the company's admittedly slow sales. "These things end up being more complex than you think up front," he said. "If anything, we may have to invest even more money to get to the goals we had than we had thought up front . . . I don't think that good virtual reality is fully there yet. It's going to take five or ten more years of development before we get to where we all want to go."

The following day, Palmer Luckey appeared in public for the first time in months to offer his own testimony to the court. He denied ZeniMax's claim that he'd violated a nondisclosure agreement, explaining that he'd used their code, with permission, to run programs on his Rift prototypes, but never revealed anything proprietary. "I didn't take confidential code," Luckey said. "I ran it and demonstrated it through the headset. It is not true I took the code."

On February 1, the trial ended with a mixed verdict for both parties. The Dallas jury ruled that Oculus did not misappropriate trade secrets and that Carmack hadn't done anything illegal. But it also found that Palmer Luckey—and his entire company, by extension—had violated a nondisclosure agreement and infringed on ZeniMax's copyrights. As a result, the jury awarded ZeniMax half a billion dollars in damages—$300 million payable from Oculus, $50 million from Luckey, and $150 million from Brendan Iribe.

Despite the financial penalties, Oculus played off the verdict as a victory—the trial proved, the company said, that the Rift was not based on stolen technology. The company also vowed to appeal the verdict, and legal wrangling was likely to continue for years to follow. But even if the decision never was overturned,

Facebook didn't seem particularly worried about the money. Later that day, the company announced its fourth-quarter financial results, saying revenue had totaled $8.8 billion for the quarter, up 51 percent from a year earlier, and well above Wall Street's expectations of $8.5 billion.

———

Two months later, on March 30, 2017, Oculus VR announced that Palmer Luckey—inventor of the Rift headset, cofounder of the company, and poster boy for virtual reality—was leaving Facebook and ending his relationship with the business that had been so closely tied to his identity.

Oculus declined to discuss the reasons for Luckey's departure but sang his praises in a statement released to the media. "Palmer will be dearly missed," it said. "Palmer's legacy extends far beyond Oculus. His inventive spirit helped kickstart the modern VR revolution and helped build an industry. We're thankful for everything he did for Oculus and VR, and we wish him all the best."

Luckey also kept quiet, maintaining the media blackout he'd imposed since October. People familiar with the company said he'd been gently pushed out or encouraged to leave on his own. Certainly all his recent troubles were a factor—loss of credibility from the Nimble America fiasco, and embarrassment and liabilities related to the ZeniMax lawsuit. But it's also clear that the company was growing out of its wild youth and was expected to become a respectable division of one of the world's biggest businesses. The future of Oculus VR wasn't sandals and board shorts, it was oxfords and suits—or at least, Facebook being Facebook, $300 sneakers and designer-made hooded sweat shirts.

In the months following his departure, Luckey kept a low profile and seemed to focus on enjoying his independence and his considerable wealth. In April, the lifelong beach bum purchased the Huntington Harbour Bay Club, a 165-berth private marina near his Orange County hometown, in a deal that cost him more than $34 million. The corporate entities used to complete the transaction were called Zeal Palace and Fiendlord's Keep—references to two locations in one of his favorite video games, Chrono Trigger.

In May, he appeared in public for the first time in months to attend Machi Asobi, an annual convention for die-hard anime fans held in Tokushima, Japan. During a cosplay event, he dressed up as Quiet, an infamously sexy female assassin from the Metal Gear video game franchise. Photos from the event show a smiling, laughing Luckey wearing a tactical harness, torn stockings, and a black vinyl bikini top.

It looked like he was having fun. But he couldn't stay away from the VR business for long. In June 2017, Luckey announced that he was investing in Revive, a start-up developing software that would allow Vive users to play otherwise exclusive content from the Oculus Store. He also announced he had founded a new VR company, though he wouldn't share any specifics—just that it was working on "some very exciting things," and they would be made for more than the Rift.

"I shouldn't be remembered as an Oculus person," he said. "Just think of me as a VR person. Sony, HTC, other companies. Everything."

Chapter 10

WE'LL USE THE ORGASMATRON

J ill the babysitter is walking across the kitchen when she notices someone sitting at the counter. "Oh my god!" she squeaks, clutching at the towel wrapped around her otherwise bare body. "Mr. Johnson! I hope you don't mind I used the shower."

Mr. Johnson doesn't respond. He is, like the babysitter, a character in an adult video, and the actor who plays him doesn't have any lines. In fact, the audience will never hear his voice or see his face, even though he does have a big part in the movie. This is virtual reality pornography, and every scene is shot from actor Preston Parker's point of view. He's standing behind the camera, and only his forearms, resting on the kitchen counter, are visible in frame.

The babysitter, played by Jill Kassidy, walks toward him and takes one of his hands. "What's wrong?" she asks, looking straight into the camera. "Where's Mrs. Johnson?"

You can imagine where it goes from there.

The first real boom in VR was porn, and adult entertainment

companies like Naughty America—producer and distributor of *Bangin' the Babysitter*—were leading the way. In just eighteen months after it produced its first VR video, the San Diego–based studio released 108 movies, making it the most prolific producer of VR content in the world. When the company operated a large booth at the annual International Consumer Electronics Show in Las Vegas in January 2017, it was the first adult business allowed to exhibit in nineteen years.

"The response has been incredible," said Andreas Hronopoulos, CEO and owner of La Touraine, Inc., Naughty America's parent company. "Our customers have embraced VR. It's just so intimate, there's just nothing else like it."

———

It should have come as no surprise that virtual reality was being used to produce pornography—the appeal of immersive 3-D video is obvious. And anyone who knows technology would have expected adult entertainment companies to be first movers in the nascent VR industry. After adult content helped popularize new media formats like VHS, Blu-ray, and streaming video, the idea that porn drives digital innovation became a widely accepted truth.

But what was surprising was how fast the VR porn boon happened and how big it had become. By the end of 2016, the new generation of commercially available VR headsets hadn't even been in stores for a year. In all of 2016, SuperData estimated that market leaders Samsung, HTC, Google, Sony, and Facebook-owned Oculus sold just over 6.1 million units worldwide. Yet even with relatively few headsets on the market, the percentage of owners who watched virtual reality porn must have been huge. Naughty America said its customers downloaded more than 20 million VR videos in December 2016 alone.

VR porn consumers were also willing to pay to view. Subscriptions to Naughty America's website grew 55 percent in 2016; customers paid $25 to access unlimited videos (including more than 7,500 traditional two-dimensional movies) for a month, or $74 for a year. And Naughty America said it converted 1 out of every 167 visitors to VR scene preview pages on its website into paying customers, compared to 1 in 1,500 for traditional scenes.

As a result, the company was making money hand over fist. Naughty America wouldn't reveal exact numbers, but said that in the eighteen months since releasing its first VR video, overall revenue had increased more than 40 percent; for all of 2016, VR-driven revenue was up 433 percent.

Naughty America was founded in 2004 as a subsidiary of La Touraine, which owned and operated a handful of adult websites with names like Tonight's Girlfriend, My Sister's Hot Friend, and My Friend's Hot Mom. Though it initially focused on producing adult movies for sale on DVD, Naughty America became the company's biggest brand after it embraced streaming Internet video.

"We try to be an innovator, a first adopter," said CIO Ian Paul. "Virtual reality had been on our radar for a long time, but always as one of those 'Oh, man, if only we could' ideas." It was only after Oculus VR successfully funded its first Rift headset via the 2012 Kickstarter campaign that Naughty America began seriously investigating the technology.

Since no one had really made VR movies before, the company

had to invent a process from scratch. "It took a lot of experimentation, a lot of investment into R&D, and basically just buying all the equipment that was available and figuring it out," Paul said. "One of the most basic problems we faced was camera placement—we originally had it too high, and it just felt weird."

The company's exact process was proprietary, but like most outfits making VR video, it was based on a relatively simple setup—two digital cameras rigged next to each other in order to capture binocular view of a scene. Postproduction involved combining the two video feeds into a single file that includes a left-eye and a right-eye perspective; when viewed through a VR headset, the images combined and appear to be a single 3-D picture.

Naughty America released its first VR movie, *Birthday Surprise,* in July 2015. "We've come a long way in a very, very short period," Paul said. "I think it's probably the same story for a lot of companies that are in this space. It's moving fast."

While Naughty America might have been producing the most VR content in the industry, it certainly wasn't the only player in the game. Adult entertainment giant Pornhub (a subsidiary of the Luxembourg-based technology and digital content conglomerate MindGeek) contracted with Rochester, New York–based virtual reality porn production company BaDoinkVR to produce videos for its websites. Smaller players included dedicated VR start-ups like VirtualRealPorn.com, as well as an increasing number of solo performers who offered one-on-one live VR videoconferencing to clients.

———

One of the most enterprising small businesses in the space was run by Ela Darling, an entrepreneur who produced—

and starred in—original virtual reality porn for her company VRTube.xxx.

Darling began her career as a pornographic actress when she was twenty-two, and shot her first virtual reality sex scene six years later, using a makeshift studio inside her college dorm room at the University of Maryland. She started it wearing an R2-D2 swimsuit and knee socks—and finished it naked, self-pleasured, and talking to the camera as if it were a real person watching.

"I'd seen some VR at school," she said, "and it was really, really cool, and I knew that it was going to be absolutely kickass for porn. But I don't come from a tech background myself. I didn't know how to do it. I figured eventually it'll be ubiquitous enough. But then I was on Reddit one day, and I saw a guy post about how he really wants to do VR porn and they've got all the tech and everything, but they just don't know how to get performers. So I reached out and we connected and shot, and it was great."

Darling was a porn actress, but also worked as a camgirl, a woman who strips her clothes off (or takes things even further) while being recorded on a webcam, for paying customers, who watch live over a streaming video connection. So she was acutely aware of the need not just to perform a sex act but to connect one-on-one with an audience member. Virtual reality offered the opportunity to break through the screen, to make a viewer feel like he was actually in the room with her, to make a real connection—emotionally, if not physically.

"It's because of the immersion," Darling explained. "You really feel like you are immersed in that space. You feel connected there, and you feel like you're in a room with a pretty girl. I actually just had a big group discussion with several of my VR cam performers this morning, and I asked them, 'What do you think is the difference between your VR customers and your 2-D customers?' And they all said that the VR people are a lot nicer. That's something I've observed too, because the way our experience works, when you go to my cam site, you're in my actual bedroom. You feel like you're actually present in my home, and when you feel like that, you're a lot less inclined to be a jerk."

"But what if that sense of immersion leads people to stop seeking out actual human contact?" I asked her. "When I talk to ordinary people about virtual reality, half the time their reaction is, 'Oh my god, VR, that's gonna turn everybody into zombies.' If we can make emotional connections and satisfy physical needs with people in the virtual world, will people stop making the effort in the real world?"

"I hear that too," Darling said. "People love to jump to this alarmist argument that people are going to stop wanting to have sex with real people, that they're not going to leave their house, they're not even going to try to have a real relationship. First of all, nobody owes the world an effort to mate with somebody else. If I decide that it would be the best experience for me to never actually date anyone and just get fulfilled in whatever ways I need, that's my decision.

"But furthermore, this is an argument that's been made for decades, if not centuries. When cult comics were really popular amongst the youth, adults were certain that they were gonna turn children into barbarians. When television started to be really popular, they thought if you spend too much time in front of a TV, then you're gonna be a mindless zombie. Then the same with the

Internet and with chat rooms. I mean, people talked about how Facebook is going to make it so you don't have real-life friends anymore, and in fact it supplements your real-life friendships. It gives you connections to people who share more than a geographic location with you.

"I think VR is going to supplement people's lives in a similar way. If I'm interested in some sex act, I can experience it in virtual reality before I'm looking down the barrel and worrying 'I don't really know how to do this.' I don't like to push the idea that it's porn's job to teach people how to have sex, but it can definitely allow people to experience a sexual situation in a safe space before they actually engage in it."

I asked her if she thought VR porn might even have a therapeutic benefit.

"Definitely. It's kind of a form of exposure therapy. If you're terrified to talk to the opposite sex and you talk to them in a virtual space, you can practice the 'being face-to-face with a girl' part, and then experience the rest of it later.

"We actually made a dating simulator that was sort of like social education," she said. "It starts off in a coffee shop, and a version of me is in front of you, and she introduces herself, and then you have the opportunity to respond with questions or comments, and then she'll respond to your question or comment based on how much of a conversation you've had together, so if you are too forward too fast, she is out of there. She does not have time for your nonsense, and it sort of teaches people that you need to be a real person, and don't go up to a girl you don't really know and just start talking about your dick, because she won't like that."

————

Why were so many adult enterprises making the leap into VR? Obviously it was for the money—according to Piper Jaffray, VR porn was set to grow into a $1 billion industry by 2020. But there were other reasons to push the tech aside from its commercial appeal. VR makes piracy harder, since the movies must be downloaded and played on specialized hardware; that means the videos can't be redistributed on the countless "tube" websites that plague the industry, pirating other companies' videos and surrounding them with ads.

In fact, while most adult entertainment companies worry about other sites taking and distributing their content, the biggest problem facing producers of VR porn was that not enough partners are willing to take the plunge. At the time, the vast majority of VR software and content was sold directly through the companies that produce VR hardware; if a consumer owned Sony's PlayStation VR headset, they could really only buy content through the PlayStation Store. And as of 2017, none of the major digital distribution services allowed adult content in their stores. Naughty America subscribers had to download files to their computer or phone and then "sideload" them into a VR video player in order to get them to work.

"A lot of these big companies are fearful of getting associated with porn," said Paul. "I think there's concern about minors accessing the content, but we've had pay-for-view on cable systems for years, so it's not like that problem can't be solved technologically. There's a way you can do verification to avoid that. So I think a lot of it is political."

The big content platforms might have done well to open up their doors. The VHS videotape standard famously eclipsed Betamax in part because Sony wouldn't allow porn companies to license its Betamax technology. If Oculus, HTC, Samsung, or Sony became the first to relent and allow adult content in their stores, it could be a big selling point for their hardware. "If you look at the history of technology, anytime anyone's ever bet against porn, they've lost," said Paul. "Of course we want adoption to happen faster just because it's our business. But it'll happen, it's just a matter of when."

In the interim, Naughty America didn't mind waiting for the rest of the world to come around. "There's always demand for our product," said Paul. "Sometimes we joke that we're like a utility, like a water company. We just have to make sure that we monetize it in a way that we can reinvest and keep producing."

Chapter 11

MAGICAL THINKING

The Plantation Pointe complex was utterly pedestrian: boxy, off-white buildings, a sprawling parking lot, a few scattered palm trees. It was not the sort of place you'd expect to find a seventy-five-foot-tall war machine from another planet.

But there it was—an All Terrain Armored Transport, better known as an AT-AT, somehow transported straight from the Galactic Empire's motor pool to the Miami metropolitan area. It was stomping around outside and I was watching it through an open door, while staying safe inside.

Or so I thought. Suddenly the whole building seemed to shake, and then the ceiling exploded. Dust and debris rained down all around me. The AT-AT had shot a hole in the roof, and when I looked up through it, I could see the combat quadruped's massive head against the bright blue sky. It peered down at me, and its twin heavy laser cannons pivoted and fired.

Then in a flash of light, everything vanished. The AT-AT was gone, the hole in the roof was gone, even the door was gone. What I'd perceived as an exterior wall was just a floor-to-ceiling curtain.

It had all been an illusion, conjured into being through the lenses of a "mixed reality" headset—the arcane invention of a start-up called Magic Leap.

Like any good magician, founder and CEO Rony Abovitz kept his cards close to his chest. Magic Leap had operated in extreme secrecy since it was founded in 2011. Only a few people got to see its technology, even fewer knew how it worked, and all of them were buried under so many nondisclosure agreements that they could barely admit the company existed.

Yet massive amounts of money were flowing down to Dania Beach, Florida, a town of 30,000 just south of Fort Lauderdale. Magic Leap had raised nearly $1.4 billion in venture capital, including $794 million in February 2016, reportedly the largest Series C round in history. Seemingly every blue-chip tech investor had a chunk of the company, including Andreessen Horowitz, Kleiner Perkins, Google, JPMorgan, Fidelity, and Alibaba—plus, there was backing from less conventional sources, such as Warner Bros. and Legendary Entertainment, the maker of films like *Godzilla* and *Jurassic World*. By the time I visited, the five-year-old company was valued at over $4.5 billion.

That cascade of money sparked strange rumors within tech circles. Magic Leap was doing something with holograms or lasers, or had invented some reality-warping machine the size of a building that would never, could never, be commercialized. The lack of hard information further fueled the whispers. Magic Leap had never released a product. It had never given a public demonstration of a product, never announced a product, never explained the proprietary "mixed reality lightfield" technology that powered its product.

But the company was finally coming out of the shadows. When I showed up in September 2016 to write a *Forbes* cover story about Magic Leap, Rony Abovitz told me that the company had spent over a billion dollars perfecting a prototype and had begun constructing manufacturing lines in Florida, ahead of a release of a consumer version of its technology. When it arrived, it could usher in a new era of computing, a next-generation interface we'd use for decades to follow. "We are building a new kind of contextual computer," Abovitz said. "We're doing something really, really different."

Inside Magic Leap's headquarters, even the office equipment did the impossible. Viewed through the company's headset, a virtual high-definition television hanging on a wall seemed perfectly normal—until it vanished. A moment later it reappeared in the middle of the room, levitating in midair. When I walked over to it and looked at it from different angles, it appeared solid; when I moved around the room, it stayed in place, just floating there.

Magic Leap's innovation wasn't just a virtual display, it was a disruption machine. The technology could affect every business that uses screens or computers and many that don't. It could kill the $120 billion market for flat-panel displays and shake the $1 trillion global consumer electronics business to its core. The applications were profound: throw out your PC, your laptop, and your mobile phone, because the computing power you need would be in your glasses, and they could make a display appear anywhere, at any size you like.

For that matter, they could make anything appear, like directions to your next meeting drawn in bright yellow arrows along the roads of your town. You could see what a couch you were thinking of buying looked like in your living room, from every conceivable angle, under every lighting condition, without bringing it home. Even the least mechanically inclined would be able to repair their automobiles, with an interactive program highlighting exactly which part needed to be replaced and alerting them if they did it wrong. And Magic Leap was positioned to profit from every interaction: not just from the hardware and software it would sell but also, one imagined, from the torrent of data it could collect, analyze—and resell.

"It's hard to think of an area that doesn't completely change," Abovitz said.

———

Virtual reality, as we already know, is an immersive computer-generated simulation. VR headsets block outside stimuli, use sight and sound to mask the real world, and make you believe you're in a different environment. But instead of using computer simulations to replace reality, what if we just made changes to it? What if we used high-definition 3-D computer graphics to drop an alien into your living room, make your carpet look like lava, or delete the couch from existence?

The people who study simulation for a living tend to describe this as computer-mediated reality, and the people trying to sell it to the public usually brand their products as augmented reality. But whatever you call it, the idea of a system that overlays digital content onto the physical world has been around for over a century. In 1901, children's book author L. Frank Baum wrote about a pair of electric spectacles in an illustrated novel called *The Master Key: An Electrical Fairy Tale*. When the book's protagonist wears the glasses, everyone he meets is "marked upon the fore-head with a letter indicating his or her character"—*G* for good, *E* for evil, *W* for wise, *F* for foolish. (He uses the device, called a Character Marker, to convince the King of England that one of his ministers is conspiring against him.)

Baum was impressively ahead of his time—other electric gadgets in the novel foretell the invention of Taser guns and live television—but it didn't take the world long to catch up with him. By the mid-twentieth century, experiments in immersive theater (like Morton Heilig's Sensorama) and the first head-mounted computer displays (like Ivan Sutherland's Sword of Damocles) were providing proof of concept for computer-mediated reality. In the 1970s and '80s, Thomas Furness's work building high-tech cockpits for the US Air Force moved the needle even closer.

Another big step forward occurred in 1990, after engineers at Boeing were tasked with replacing the sprawling wiring diagrams that cluttered the aerospace company's factory. Workers needed reference examples when they were assembling complicated aircraft parts, but the systems were so complex they had to be set up on sheets of plywood the size of billboards, and the process was so painstaking it caused slowdowns every time the workers advanced a step in the manufacturing process. To solve the problem, researchers David Mizell and Tom Caudell proposed a head-mounted display that overlaid digitized schematics onto a reusable board, showing the wearer exactly where to lay out wires and slashing the amount of time it took to set up each diagram. Caudell called the system augmented reality, and the terminology stuck.

Boeing never deployed the system, but it influenced the work of other researchers. In 1991, a group of Columbia University computer scientists led by professor Steve Feiner started development of a prototype they called KARMA, or Knowledge-based Augmented Reality for Maintenance Assistance. "A well-designed virtual reality can make it possible for one or more suitably outfitted users to experience a world that does not exist," Feiner and his team wrote about the project in an academic journal. "There are many situations, however, in which we would like to interact with the surrounding real world. An augmented reality can make this possible by presenting a virtual world that enriches, rather than replaces, the real world . . . this approach annotates reality to provide valuable information, such as descriptions of important features or instructions for performing complementary tasks."

The KARMA prototype, built in Columbia's Computer Graphics and User Interfaces Lab, was an extraordinary tech support system for an ordinary office laser printer. Feiner's team attached ultrasonic tracking devices to key components of the printer so KARMA could monitor its exact location and orientation. Users would wear a custom head-mounted display with a see-through screen that displayed low-resolution computer graphics. When they looked at the printer, KARMA would draw red lines around the printer's paper tray, fainter "ghosted" lines showing where the tray would be if it was opened, and a pulsing arrow indicating that the user should pull it out. "We have developed a preliminary set of rules that allow us to augment the user's view of the world with additional information that supports the performance of simple tasks," Feiner wrote.

Other labs focused on developing augmented reality systems that allowed users to complete tasks at a distance. Researchers had noted that it was difficult to operate remote-controlled devices accurately, since the operator worked on a time delay and lacked normal visual and physical feedback. It's harder to drive a car via remote than to operate it from the driver's seat, for instance. So in 1992, Louis Rosenberg, a Stanford University graduate student working at the US Air Force's Armstrong Laboratory, designed a simple task—guide a robot to put a peg in a hole—and built a special AR rig to complete it.

The user strapped into an upper-body exoskeleton connected to a pair of remote-controlled robotic arms. When the user moved his or her arms, the robots matched that movement. Users also wore a headset that made it appear as though the robot arms and the pegboard were directly in front of them. Finally, Rosenberg installed a series of fixtures in front of the user that allowed them to hold an actual peg and insert it into an actual hole, simulating what they wanted the robot arms to do at a distance. The effect of the system was to give users an illusion of presence at the pegboard and to receive tactile feedback on the task as they completed it. As a result, they were able to complete the job with less difficulty, and Rosenberg discovered that an AR system could significantly improve operator performance.

Over the next few years, augmented reality went prime time. In 1995, a company called Princeton Video Image inserted computer-generated advertisements into the live television broadcast of a Minor League Baseball game, making them appear to be billboards on the backstop behind home plate. The practice quickly became commonplace at all types of live sporting events. In January 1996, the Fox network's broadcast of the National Hockey League All-Star Game debuted a more controversial piece of AR technology, called FoxTrax, which gave the puck a blue glow and a comet's tail so it was easier to see on TV. Hockey purists hated the cartoonish effect; it was abandoned after Fox lost the rights to broadcast NHL games two years later, but is still remembered as one of the worst innovations in sports history.

Other AR implementations were embraced by fans. Stanley Honey, an engineer who led the FoxTrax initiative while working as executive vice president of technology for News Corp., spun that technology out into a new start-up called Sportvision and, in September 1998, during an ESPN broadcast of a National Football League game, debuted a new system called 1st & Ten. Viewers tuning in to watch the Cincinnati Bengals play the Baltimore Ravens were surprised to see a yellow first-down line that seemed to be painted across the grass of Ravens Stadium, and even disappeared underneath players who crossed it, but somehow moved downfield as play progressed.

Sportvision achieved the effect by placing sensors on ESPN's cameras that captured a variety of data, from the attitude of the tripod to how much the lens was zoomed, and used that to create a computer model of where the field was and where the first down line should appear—just like KARMA knew where the laser printer was and where its paper tray should be. A complex color-matching system made sure the system drew the line only over green grass, and not on top of players or the ball. Later that year, other networks followed suit with similar systems, and today augmented reality displays are commonplace in professional sports, ranging from Major League Baseball's pitch-tracking system to virtual lane markers in Olympic swimming and skating events.

AR went mobile in 1997, when Steve Feiner's team at Columbia hacked together a computer and a GPS system, and loaded it into a backpack wired to a head-mounted display. They called it the Touring Machine, and when users wore the rig and walked around campus, the system superimposed building names onto their view of the world, and flagged points of historical interest. In 2000, a team working out of the Wearable Computer Lab at the University of South Australia took a similar system and turned it into the first augmented reality video game, ARQuake. Based on a popular first-person shooter, the game was heavily modified to run on a head-mounted display and receive control information from expensive surveying-grade positioning sensors. As players walked around in the headset, it overlaid monsters and obstacles onto their physical environment.

But AR didn't hit the mainstream until camera phones became commonplace, since they combined the ability to record live video with the computing power necessary to process it. In 2008, an Austrian company called Wikitude launched AR Travel Guide, the first location-based augmented reality app. "Users may hold the phone's camera against a spectacular mountain range and see the names and heights displayed," the company bragged on its website. In 2009, the magazine *Esquire* published an issue with several augmented reality features and billed it as "the first-ever living, breathing, moving, talking magazine." The cover featured the actor Robert Downey Jr. sitting astride a QR code, and when users scanned it with their phones, he jumped off the page to describe the tech and show a clip from his upcoming movie *Sherlock Holmes*.

Google dipped its toes into AR in 2013 when it released Project Glass, a pair of spectacles that made a virtual computer screen appear to float in front of the user. Fewer than 10,000 units were distributed before privacy and safety concerns led Google to shelve the project. But in the summer of 2016, a Google spin-off company had better luck. In July, mobile application developer Niantic released Pokémon Go, a game that used smartphone cameras to make animated monsters appear to exist in the real world—or at least on the screen of a phone. The app debuted on top of the download charts and surpassed $500 million in revenue in just over sixty days.

But for all its success, Pokémon Go was a pretty limited example of augmented reality in action: the app could make a Pikachu appear in your living room, but the creature didn't know it was there.

———

That's where Magic Leap came in. CEO Rony Abovitz described the company's technology as mixed reality—digital content that interacts with the physical world. If augmented reality can make a Pikachu appear in your living room, mixed reality makes it come to life. An MR Pokémon could recognize its environment and interact with real-world objects; it might escape capture by ducking behind your couch, or even by turning off the lights so it can hide in the dark.

"One of the reasons we call it mixed reality is because it's taking the analog world as you know it and then blending in the digital all around you," Abovitz said. "It's a very different experience from chasing Pokémon on a phone, a very different experience from a VR headset. It should feel very natural and shouldn't feel confined; you shouldn't feel claustrophobic."

How does it do it? The centerpiece of Magic Leap's technology was a head-mounted display, but Abovitz told me the final product would fit into a pair of spectacles. When I was wearing the prototype, an array of optics sat in front of my eyes, but didn't block my view of the world; instead of drawing images on an opaque LED display, Magic Leap's hardware projected an image using an optics system built into a piece of semitransparent glass.

This photonic lightfield chip was Magic Leap's most proprietary and secret technology, so the company wouldn't say exactly how it works. But the basic idea is kind of like an old-fashioned TV. In a cathode ray tube, a stream of electrons scan across coated glass, excite phosphors that emit light, and draw a picture on the screen. In Magic Leap's system, a stream of photons scan across the lightfield chip, bounce off nanoscale structures that reflect light, and draw a picture directly on your retina.

"The idea is that we are simulating a signal that your brain has evolved to process," Abovitz said. "What Magic Leap is doing is trying to get out of the way." Instead of creating an image on a screen, the photon chip feeds your eyes information and lets your brain create the picture, he explained.

The result is incredibly realistic virtual images. Since the photonic chip sends light directly into a viewer's eye, it can display images at a much higher resolution than a physical screen can; it isn't limited by the size of components like light-emitting diodes. The chip also eliminates many of the comfort issues associated with 3-D displays. With Magic Leap's system, your eyes focus on the point where an object appears to be, not on the screen where it's actually displayed. That makes 3-D simulations less taxing on your eyes—and more convincing to your brain.

But there's much more to this technology than an advanced 3-D display. Mixed reality is interactive, so the headset is constantly gathering information about the world around it, scanning the room for obstacles, listening for voices, tracking eye movements, and watching hands.

In one of its demos, the Magic Leap team showed off a computer-generated "virtual interactive human," life-size and surprisingly realistic. Abovitz and his team imagined virtual people (or animals or anything else) as digital assistants—think Siri on steroids, except with a physical presence that makes her easier to work with and harder to ignore. Ask your virtual assistant to deliver a message to a coworker and it might walk out of your office, reappear beside your colleague's desk via his or her own MR headset, and deliver the message in person.

In a mixed reality world, computing power isn't confined to a gadget on your desk. It's something that you can link to any object, real or virtual, giving it awareness of its location, intelligence about its purpose, and insight on how you might want to use it. "Think of it as the future state of computing," Abovitz said, "where the world is your desktop." First we had mainframes, then PCs, then mobile devices. If Magic Leap had its way, the next generation would be virtual.

"This is not about entertainment or just playing video games," said Thomas Tull, the billionaire founder of Legendary Entertainment. "This is a different way of interacting with the world, a new generation of computers. I think this will end up being a very, very important company."

———

Rony Abovitz had always surrounded himself with visions of a high-tech future. Born in 1971 to Israeli immigrants living in Cleveland, he grew up fascinated with computers and science fiction. "My generation are the children of Steve Jobs and George Lucas," he said. "We grew up on that, and our brains all got scrambled . . . My friends and I all wanted to be Luke Skywalker and defeat the Death Star and build C-3PO."

When he was eleven, the family moved to South Florida; he started high school at thirteen, a year early. After graduation, he was accepted to MIT but chose the University of Miami, staying close to home. He received a bachelor's degree in mechanical engineering in 1994 and a master's in biomedical engineering two years later—and then he started thinking about *Star Wars* again.

Abovitz cofounded his first company, Z-KAT, in 1997. "I decided to build the medical droid from *Star Wars* because I thought—I mean, this was the genuine impulse—I couldn't build an X-wing fighter, because that wasn't something I could explain to my parents after graduating," he said. In 2004, Abovitz and several of his cofounders spun out Z-KAT's robotics group into a new company, Mako Surgical, which made robotic arms to assist doctors in orthopedic surgeries. Demand for the droids was high, and in 2008 the company went public, raising $51 million.

Working full-time at Mako and married with a young daughter, Abovitz found an outlet for his creative side in a project he called Hour Blue—his own fictional fantasy world, an alien planet full of fantastic creatures like talking robots and flying whales. In 2010, he started a new company, Magic Leap Studios, to develop the project as a graphic novel series and a feature film franchise.

"I was the only employee, and it was literally in my garage," Abovitz said. "My mom made a piece of canvas with some colored letters on it that said, 'Magic Leap Studios.'" For help on the project, he used some of the cash he'd banked from Mako to hire Weta

Workshop, the New Zealand–based special effects and creative-development shop best known for its work on *The Lord of the Rings* film trilogy, to develop imagery based on his story ideas and flesh out the world. Meanwhile, inspired by sci-fi novels like William Gibson's *Neuromancer* and Vernor Vinge's *Rainbows End*, Abovitz became frustrated that the augmented and virtual reality he'd read about in fiction wasn't available in the real world and began thinking about how to make it so.

"It was a unique moment . . . Reality and science fiction began to merge," said Richard Taylor, CEO of Weta Workshop and a member of Magic Leap's board. "The fictitious technologies that we were creating for Hour Blue were developing in parallel to the real-world applications of augmented technology that Rony was beginning to explore."

In 2011, Magic Leap Studios changed focus and became Magic Leap, Inc., with Abovitz hiring a small team to help him develop this idea of mixed reality. Before long, the company had working prototypes.

"The first time we had a single pixel in space, where we could move it around the room, we were so excited," Abovitz said. "Other people were like, 'What is that? That's just a dot.' But we knew. I knew at that point this was going to work."

He also knew he was going to need a lot more money. Abovitz had initially funded the company with his proceeds from Mako's IPO. After Mako was acquired in 2013 by medical device manufacturer Stryker Corporation for $1.7 billion, he invested some of those proceeds as well. Abovitz wouldn't reveal the exact amount he spent to get the company running ("in the millions" was all he'd say), but he knew it wouldn't be nearly enough.

Fortunately, the technology sold itself. "When we'd talk to people about what we were doing, they didn't believe us," Abovitz said. "Then they'd fly in and go, 'Oh . . . you've actually made these things happen.' That was the dynamic of everyone who invested; [they went from] 'This is impossible' to 'We want in.'"

In February 2014, Magic Leap announced it had raised more than $50 million in seed funding from private investors. Eight months later, the company closed a Google-led $542 million Series B funding round.

"We invested in Magic Leap because we believe their light-field technology is the next big inflection point in technology after the PC, the Web, and the smartphone," said James Joaquin, co-founder of Obvious Ventures, a San Francisco–based venture capital firm. "It has the potential to transform multiple global economic categories, including entertainment, education, and productivity."

Magic Leap's record-breaking $794 million Series C round, announced in February 2016, was led by Chinese technology conglomerate Alibaba and included follow-on investments from Google and Qualcomm Ventures, plus new money from Fidelity Investments, JPMorgan, Morgan Stanley, and T. Rowe Price. "I don't think it would be mischaracterizing it to say there was a frenzy," said Thomas Tull. "This company has had absolutely no problem attracting blue-chip investors, and it's extraordinarily well capitalized."

———

Rony Abovitz didn't seem like the captain-of-industry type—unless your model mogul was Willy Wonka. Abovitz lit up like Roald Dahl's genius chocolatier as he led me on a tour through

Magic Leap's new headquarters in Plantation, Florida, a campus more befitting his vision than the nondescript offices fifteen minutes away in Dania Beach. He pointed out "cool" machines, admired tools, and encouraged me to climb a ladder so I could check out high-end air filters in the ductwork. He was friendly and cheery, casual in speech and attire. On the few days I visited the company, his employees commented on what a nice guy he was as often as they mentioned his intellect.

He was also liable to get lost in his work. On one Friday afternoon, he was missing in action a half hour after a tour of the new headquarters was scheduled to start—a not uncommon delay but a problematic one, since Abovitz comes from an Orthodox Jewish background and was planning to leave work early to observe the Sabbath. Eventually one of his executives discovered he had been on-site the whole time, sitting in his car in the parking lot, absentmindedly engrossed in a telephone call.

Magic Leap broke ground on the new 259,000-square-foot campus in October 2015, and when I visited a year later, Abovitz expected the majority of its 850 employees to move in before the end of 2016. The rest of its workers were scattered in nine offices around the globe, not just in tech hot spots like Silicon Valley and Austin but also in far-flung outposts like Wellington, New Zealand, and Tel Aviv. Some groups were already in place at the new site, including a machine shop and several engineering teams. It was important to Abovitz to keep critical development teams together as part of an "agile hardware" model, which allowed the company to produce "literally many hundreds of iterations" of its headset prototype. "Part of why Magic Leap can iterate so fast is because we have all the right people in the right places," Abovitz said.

The company was also building manufacturing facilities on the Plantation campus. "This is the most spaceshippy part of Magic Leap," Abovitz said as he led me into the production line areas: a series of long self-contained modular bays lined up like submarines in a port. Each line could be activated as needed, ramping up production from thousands of units a year to more than a million.

Abovitz wanted Magic Leap to stay in Florida. One of the benefits of manufacturing there was that it allowed the company to keep its secrets. If it had been headquartered in Northern California, that would be nearly impossible, given Silicon Valley's job-hopping culture and well-lubricated rumor mill. Of course, Magic Leap would also have an easier time hiring, but so far the company's technology had been a strong draw, pulling people in from the Valley and other tech hubs. "We're bringing a crazy amount of very capable high-end engineering and manufacturing tech talent into Florida," he said.

So now that it had a production line, what was Magic Leap's timing for entering the market? I asked. Abovitz was tight-lipped. "Soon-ish," he said. He was also reticent about what the headset will cost. "Not a luxury product," he finally offered.

Magic Leap was developing much of its content in-house, and had already hired several well-known video game designers, cartoonists, artists, and writers. Neal Stephenson, author of *Snow Crash*, the seminal 1992 novel about virtual reality, was working as Magic Leap's chief futurist and developing an undisclosed game. Other content was coming from partners like Weta Workshop, which operated a twenty-five-person lab with Magic Leap in New Zealand. Their first project, Dr. Grordbort's Invaders, was an action-oriented game set in a steampunk version of a future British empire: the player used a ray gun to fight off evil robots that seem to break through the walls of their living room and fly around the house.

Another high-profile partner allowed Magic Leap to work with characters that were close to Abovitz's heart. In June 2016, the company announced a strategic partnership with ILMxLAB, the immersive entertainment division of Lucasfilm Ltd., and opened a joint research lab on Lucasfilm's San Francisco campus. By the time I visited the Florida headquarters, the collaboration had already produced several mixed reality experiences set in the *Star Wars* universe, including the AT-AT attack that appeared to blow open a section of the building.

"It's like being in the earliest days of cinema, where people were sort of inching their way along in terms of understanding the format and what was possible and what was compelling," said Vicki Dobbs Beck, the head of ILMxLAB.

But entertainment was only the first part of Magic Leap's plan—the foot in the door. Business applications would follow. Medical systems were likely to be a sweet spot, like 3-D radiology displays that float above patients in an operating room. Retail would be an early target too. "People love to go to a mall or store and see the things they're going to buy in the flesh," Abovitz said. "I think we can bring some aspects of that to your home that are impossible on your phone or tablet or TV." Imagine a "try it on" button on your favorite online clothing store—click it and then look in a mirror, and the headset renders the outfit on your body so you can see how it looks.

———

Naturally, Abovitz wasn't the only entrepreneur who saw opportunity in this field—all the big names in tech were scrambling to get a piece of the action.

At the time of my site visit, Magic Leap's biggest competitor was Microsoft, which announced an augmented reality headset called HoloLens in 2014. A preproduction version of the hardware, the HoloLens Development Edition, had shipped to an unspecified number of Microsoft hardware and software developers in March 2016. Google had already dipped its toes into AR in 2013 with its ill-fated Google Glass, but its investment in Magic Leap indicated ongoing interest. Apple was working on AR as well, but it was unclear whether they were developing their own headset or adding capabilities to their iPhone line. And then there were a few other well-financed start-ups, most notably San Mateo, California's Meta (which had raised $50 million in funding) and Mountain View, California based Atheer (with $23 million raised), that were working on their own AR headsets.

Magic Leap wouldn't be the first or only company on the market. But its technology still seemed light-years ahead. Its success so far reminded me of another time a group of hardworking visionaries built something in Florida that rocked the world—and more than one Magic Leap employee compared what they were building in Plantation to NASA's early days in Cape Canaveral.

"To me, this feels like we're crawling the rocket out to the pad," Abovitz said.

Epilogue

THE AGE OF
THE UNREAL

uring the years of reporting covered in the pages of this book, on perhaps a dozen occasions I experienced something I've come to describe as a "dinosaur in a hotel room" moment. It's something that had never happened to me before, and something that's hard to explain because it's new and profound. But I think it's something I'm starting to understand. I know other people have experienced it recently. And I suspect the rest of humanity is going to have to deal with it soon.

Every time I've experienced one of these moments, I've had some kind of high-tech hardware strapped to my head. But the dinosaur in the hotel room is bigger than the thrill I get from seeing a convincing simulation, and stranger than getting an acute stress response just because a virtual *Tyrannosaurus rex* roared in my face.

It's the moment that comes just after that—a moment of conscious recognition that reality is not as it seems. When it happens, I know I'm in a simulation, I know what I've been experiencing isn't actually real . . . but I simultaneously know that *it is real*, as real as anything I'll experience after I take off the headset. I saw it, I felt it, I believed it, I'll remember it just as vividly as any other memory. It happened. It doesn't matter how.

By the time this book is published, more than two years will have passed since a new generation—the first really good generation—of virtual reality headsets arrived in stores. From a business perspective, the Oculus Rift, the HTC Vive, the Play-StationVR, and all the others have been moderate successes. Smart people who track these things tell me that at the end of 2017, the VR and AR market was already worth more than $11 billion, and that it will be worth more than $200 billion in 2021. Not bad for an industry that barely existed five years earlier.

Still, despite all the hype and promise, VR hasn't changed the world, and it's not likely to very soon. The industry and the technology are still in their infancy. The success of companies like Oculus and Magic Leap could be cut short for countless reasons—anything from shareholders demanding better returns to the global economy sinking into a depression. And some of the most influential tech companies in the business—Apple, most notably—still haven't really entered the market.

But I can't ignore a dinosaur in a hotel room. Those moments have altered my perception of reality, convinced me it's bigger and weirder than ever before. It's like I've discovered another plane of existence, stepped through a wardrobe to find a world where animals talk and magic is real. I can't imagine that people won't want to explore that, can't believe this technology is a fad that is going to go away. It's too big, too profound; it's going to allow us to experience things that used to be impossible, unreachable, or unthinkable. For better or worse.

So yeah, VR hasn't changed the world yet. But I can imagine how it will.

———

Henry Dorsett hated the sound of his alarm, but the high-pitched screech was the only thing that could wake him up in the morning. Even on a good morning he'd slumber straight through anything short of a siren, and this wasn't a good morning. He'd stayed up way too late playing poker with a couple of fascinating guys who worked for a solar power start-up in Liaoning Province—and that was after staying up late the night before that in order to attend a concert at the Acropolis.

He squeezed his pillow tight over his head to drown out the noise, but eventually the alarm won and Henry reached over to silence it. His glasses were on the bedside table, so he sat up, put them on, and focused his bleary eyes on the bright yellow words that had appeared out of nowhere, floating above the foot of the bed.

Good morning! they read, next to a smiling emoji. Below them, there was a smaller line of text: Monday, January 5 | Sunny, 12°C.

Henry groaned when he climbed out of bed, and as he walked out of the bedroom, the message remained stationary until it passed out of his peripheral vision and blinked out of existence.

Down the hall and into the kitchen for breakfast. "News," he said, his voice still thick with sleep. A nine-inch-tall cartoon fairy—the currently selected appearance for Miri, his personal digital assistant—popped into the room in a flash of light and hovered over the countertop.

"There was a magnitude 6.1 earthquake in the Gulf of California this morning," she said in an Irish accent. "No fatalities have been reported and your mother's house was not damaged." The sound was projected from a conductive speaker in the temple of his glasses, but stereo audio tricks made it seem to emanate from the fairy, even as Henry puttered around the kitchen making his breakfast.

"The Jets lost to the Patriots, thirteen to twenty-seven," Miri continued. "Would you like to experience the highlights?"

Henry flicked his hand at the fairy, and it vanished in a burst of glitter.

While virtual reality has been getting most of the atten-tion, augmented reality may turn out to be far more commonplace and a much bigger market. As companies like Magic Leap perfect and shrink their products, AR glasses will combine the functionality of smartphones and computers, seamlessly inserting digital information all around us. They'll replace physical screens—no more monitor-hogging space on your desk, just a virtual desktop floating in air that you can dismiss with a gesture. And they'll enable smart personal assistants that know your habits and interests and interact with you like they were real people. Analysts say the AR market could approach 3.5 billion installed devices and $90 billion in revenue by 2020.

After breakfast, Henry showered and then put on his work clothes—a Yomiuri Giants T-shirt and a pair of old sweatpants. As he was tying his shoes, Miri blinked back into existence, her little wings beating furiously as she flew a tight circle around the inside of his closet.

"Your prototyping meeting is scheduled to begin in five minutes," she said, "but Alice and Bob have already logged on from Raleigh and Antwerp, respectively." Henry nodded and walked down the hall to his office.

The room was mostly empty, decorated with posters and pictures on the walls, but the only furniture was a small table and a single chair. Henry took off his glasses, put them on the table, and picked up a pair of fabric gloves embedded with small tubes and sensors. He pulled them on and then picked up his VR headset—a sleek glass-and-plastic contraption, like a pair of ski goggles, that fit snugly over his eyes and ears, blocking outside sensory input.

He stood in darkness for a moment before the interface started up, and Miri appeared, hovering in the void a few feet away from him.

"Would you like to enter the meeting now?" she asked him.

"Yes, please."

———

The first-generation Rift and Vive virtual reality head- sets already produce surprisingly realistic and immersive simulations, and the next few iterations are only going to get better. Whatever device we're using in five, ten, or twenty years, it's likely to be lightweight, comfortable, and utterly convincing. Input devices will evolve along with them. While hand tracking and voice recognition will allow us to do most of our work without actually using any physical controllers, I suspect we'll sometimes wear data gloves in order to provide the most accurate tracking and tactile feedback. Imagine a glove that could stiffen and push back against your fingers to actually make it feel like you're holding a virtual object.

———

Miri snapped her tiny fingers, and in an instant Henry was standing in the middle of an open city square, surrounded by marble buildings decorated with Romanesque carvings. A few feet away, a man in a dapper gray suit was chatting with a woman in a navy double-breasted jacket and trousers. They turned toward him, and the woman waved and smiled.

"Hi, Alice. Hi, Bob," Henry said.

———

Virtual reality frees its users from the constraints of physical presence. This technology will connect people who live and work in different places better than any video chat or telephone conference call. VR chat-room software can already track its users' movements and replicate them in a virtual space. Future versions of the technology will capture the most minute details of their expressions and gaze, creating lifelike simulacra that actually look, act, and feel like the people they represent. That means virtual meetings where you can look in your coworkers' eyes, see if they're smiling or paying attention, follow their gestures, and better understand their intent.

And since it's not restricted by the boundaries of geography, VR will allow us to use any environment as a meeting place. Maybe a formal business meeting calls for a corporate boardroom, but social events could take place on a beach, on the surface of Mars, or in a picturesque city square. A sufficiently advanced simulation could even use live data from video cameras to create real-time replicas of an actual place. Just imagine if city engineers studying traffic patterns could stand in the middle of the Beijing–Hong Kong–Macau Expressway during a severe traffic jam.

———

Henry glanced around the plaza. "You moved the meeting to Venice, huh?" he said. "And you're wearing suits. I feel like I'm underdressed." Glancing down, he gestured at the designer jeans and white button-down shirt he'd assigned to his standard workday avatar.

"Just for fun," Alice said with a shrug. "And I like your outfit. Is that a new hairstyle too?"

Henry reached up and tousled his avatar's shaggy red curls. "Sure is," he said. "When you're bald in real life, it's always nice to pretend."

———

Simulating locations will allow us to go anywhere, and simulating people will allow us to be anyone. Virtual reality untethers people from their bodies, allowing them to change their appearance at will. This can be used for minor cosmetic changes—like changing clothes or the style of your hair—or bigger flights of fancy, like giving yourself wings, tentacles, or whatever you can imagine. Or it can lead to something far more profound. VR users can change their real appearance to reflect their ideal identity, to avoid persecution or prejudice, or simply to stay anonymous. As VR chat software becomes commonplace, it may drive a thriving market in selling avatar clothing, accessories, and designs;

micropayments will buy access to countless upgrades from both independent and world-renowned designers.

———

Henry touched his thumb to his index finger, and a small menu, like an artist's palette, sprung into existence in his hand. "Should we get to work?" he asked. He touched an icon on the menu with his right index finger, and a 3-D model of a small machine appeared in front of him, floating in the air.

"This is the latest render of the design," he said, reaching out and grasping the object with his fists on both sides. "It's much better, but if you look at the cooling unit, there are still problems." Henry spread his arms, pulling the 3-D model apart so that its components were all visible, like an exploded-view drawing. Then he released his grip, reached into the cloud, and grabbed one particular part. Grasping it on either side with two fingers, he pulled and expanded the model until the component was large enough to examine in detail.

———

Because virtual reality allows us to model anything— from microscopic machine parts to entire buildings—it's going to be an invaluable business tool. Even now, companies around the world are using VR headsets at work. Engineers in Ford's Virtual Reality Immersion Lab use the Oculus Rift to examine 3-D models of new car designs. Consumer products manufacturer Kimberly-Clark Corporation uses VR to create virtual retail spaces, allowing them to test advertising and merchandising concepts without having to send anything to brick-and-mortar stores. NewYork-Presbyterian Hospital is using the technology to create relaxation programs that help burn patients manage their pain. Like computers did in the twentieth century, AR and VR will spread into every industry, and change the way the world works.

———

After the team finished examining the virtual prototype and came up with a list of changes to design into the next model, Henry swiped at the virtual components with the back of his hand, and they faded away, leaving the three of them standing alone again in the Venetian plaza.

"We're in good shape, then. I'll pass these notes along to engineering," Henry said. "Let's meet again same time on Monday. Have a good weekend, guys."

"You too," Alice said. "Any interesting plans?"

———

Since video game enthusiasts were largely responsible for designing, funding, and purchasing the first modern wave of VR headsets—and because hard-core PC gamers are among the few people who own computers powerful enough to support them—VR will make its first appearance in most people's homes as a high-end gaming machine.

And even now, the games people play are surprisingly great. I'm particularly enamored of a multiplayer game called Star Trek Bridge Crew, which allows a group of players to take roles on the bridge of a starship and pilot it through various combats and crises. It's exciting and fun and draws you into its fictional world like few other video games can do.

And personally, I can't wait to play virtual reality tabletop games, to sit down at a common table with friends in places all over the country, face-to-face, for a night of Scrabble or Monopoly or Dungeons & Dragons. Imagine the animated interactive experiences—like a game of Risk where the board has weather patterns and all the battles are animated out in 3-D glory.

And the possibilities for leisure go far beyond gaming. It's already possible to watch live sports and concerts in virtual reality. Someday soon VR chat rooms could join bars and nightclubs as a fashionable place to hang out and socialize.

Virtual reality can be used for self-improvement time too. I want a VR coach to teach me how to fence. Or a simulated Bruce Lee to teach me Jeet Kune Do.

———

Henry spent his weekend hiking in the mountains—the real *mountains, out in the wilderness, not a simulated version. He tried to get out of the house and into nature as often as possible, since he knew his VR rig made it all too easy to never leave home. Virtual shut-ins were common these days, and "going native" in VR was increasingly accepted as a normal way to live your life; since 24-7 users could still socialize, work, and explore the world, most folks figured there was no harm done.*

But Henry was old enough to remember when spending all your time online was considered a real problem. And besides, he thought, there were some things that virtual reality just couldn't reproduce— the feeling of the wind on your face, the smell of fresh flowers, the serendipitous moments of beauty that could never fit into a program.

In these early days of virtual reality, some people remain cynical about the technology, and rightly so. There are real dangers of society becoming alienated, of VR users choosing an attractive fantasy over the messy details of real life. We'll have to learn how to moderate our exposure, and how to help those who disappear into VR to the detriment of their physical and mental health.

For now, maybe we should keep VR away from children. Psychologists say that kids under eight lack the faculties to understand the difference between what's real and what's not, and even teenagers have trouble regulating their emotions when confronted with an imaginary or simulated problem. We should take care not to overexpose them until we fully understand how VR affects developing minds.

But even though I acknowledge the dangers, I still think the benefits of virtual worlds will outweigh the problems. Besides, many cynics have the luxury of privilege: they might not worry about becoming a shut-in if they were disabled and looking at VR as a way to explore the world; or they wouldn't fret about protecting children if they were an impoverished kid in a developing country eager to learn in a virtual schoolroom. (Of course, this supposes that we're able to get the currently way-too-expensive VR hardware into the hands of those people who might need it most—another problem that we'll have to overcome.)

And then there are the critics who say VR is hollow, that it never lives up to the real world. For now, that's absolutely true. I might be able to pop on a headset and experience what it's like in the stands of a live sporting event, but I know I won't have as much fun as if I were there in person. We use all our senses, not just sight and hearing, and right now that's all that VR can convincingly produce.

That will change. Today we use headsets to access virtual reality; tomorrow we might use high-tech contact lenses or even ocular implants; someday, we may create direct brain-to-computer interfaces that render the virtual absolutely indistinguishable from the real world.

———

Virtual and augmented reality haven't changed the world yet. We're still on a gentle upward slope of the technology adoption life cycle, the time when people who buy a new product are innovators and early adopters. It's a lot like the late 1970s, when personal computers were still a cutting-edge product: back then, when people thought about computers, most of them pictured the expensive mainframes used by businesses and militaries—mysterious high-tech towers that filled up a room and were used to calculate inscrutable mathematical problems. Personal computers were a hobbyist thing. Just a geeky toy.

But a generation of inventors and innovators changed that. They created computers that people wanted on a desk in their home, designed software that made them easy to use, and built applications that made them indispensable. And today computers are everywhere—not just in homes and businesses but on our wrists and in our pockets. They're so commonplace that we don't even think about them. They're just part of the environment.

VR will undergo a similar transformation. Today, most people have read about it, talked about it, heard it's really exciting, said they want to try it. But they don't own a VR rig yet. That's a hobby-ist thing. Just a geeky toy.

But before long, this exotic technology will be mundane. We'll spend time in virtual worlds just as casually as we play video games or surf the Internet. The day is coming, and when it arrives, reality will never be the same again.

Acknowledgments

I am thankful for the help of the following people:

Brant Rumble edited this book for Blue Rider Press. Brant Rumble is a steely-eyed missileman. Brant Rumble is patient and kind. Brant Rumble knows no fear. Brant Rumble is a giant among men. Brant Rumble deserves lots of money, much love, and a stiff drink.

Brent Howard completed the book for Dutton, and I am immensely grateful to him for taking on the project and seeing it to completion.

Chris Parris-Lamb is my agent, and one of the smartest people in the business. I am lucky to know him and I am grateful for his insight and support. Thanks as well to everyone at the Gernert Company for all their hard work.

Thank you to everyone at Penguin Random House, including Christine Ball, John Parsley, Susan Schwartz, Joel Breuklander, Amanda Walker, Beth Parker, Carrie Swetonic, and Cassidy Sachs. And thank you to the Blue Rider crew, including David Rosenthal,

ACKNOWLEDGMENTS

Aileen Boyle, Jason Booher, Kayleigh George, Gwyneth Stansfield, Terezia Cicelova, and Linda Cowen.

I am grateful to my editors at *Forbes*, Randall Lane and Michael Noer, for sharing their wisdom, skill, and guidance. And I am grateful to my very first editor, Garrison Hoffman, who welcomed me into the college newspaper that became my second home, and who always encouraged my work. He was a good friend and will be missed.

Thanks to my parents, Barbara and Larry Ewalt, for everything. Thanks to my nieces and nephew, Casey, Madeleine, Sophia, and Sid, for being my favorite people. Thanks to Elissa, Indrajit, Jeff, Kathy, Mal, Jocelyn, Eric, and all my kith and kin, for their support. Thanks to Michael, Jessica, and Marian for being my Brooklyn family.

And most of all, thanks to my wife, Kara, who put up with a wholly unreasonable amount of stress while I worked on this project, and somehow didn't kick me out of the house. She kept me healthy and sane and gave me a reason to keep going. In the immortal words of Homer Simpson: "We have a profound, mystical understanding . . . in your face, space coyote!"

Notes

For the purposes of storytelling, some of the conversations and interview sources described in this book have been edited for length and clarity, but aside from those changes, quoted dialogue is presented as it was spoken. Portions of this manuscript were originally published by the author in *Forbes* magazine or on Forbes .com and are used here with permission.

The sources for unique facts or quotes that are not cited within the text of this book are as follows.

CHAPTER 1—PYGMALION'S SPECTACLES

17 ***"subterranean cyberspaces"***: Howard Rheingold, *Virtual Reality: The Revolutionary Technology of Computer-Generated Artificial Worlds—and How It Promises to Transform Society*. New York: Simon & Schuster, 1992, 379–80.

18 ***Gu Hongzhong's eleven-foot-wide* The Night Revels:** "The Night Revels of Han Xizai | Digital Scrolling Paintings Project." Scrolls.uchicago .edu. N.p., 2017. Web. Accessed January 9, 2017.

18 ***the Irish-born artist Robert Barker***: "Lost Edinburgh: Calton Hill and the Invention of the Panorama." https://www.scotsman.com/lifestyle /lost-edinburgh-calton-hill-and-the-invention-of-the-panorama-1 -4032338. Accessed January 9, 2017; *Panorama of Edinburgh from Calton*

Hill by Robert Barker (1739–1806) (GB-237—GB237 Coll-1092)—Archives Portal Europe. Archivesportaleurope.net. Accessed January 9, 2017.

19 *a building in Leicester Square*: Kathryn Kane, "Robert Barker's Leicester Square Panorama: The Rotunda." The Regency Redingote, August 3, 2012 https://regencyredingote.wordpress.com/2012/08/03/robert-barkers-leicester-square-panorama-the-rotunda. Accessed January 9, 2017.

20 *view the world in three dimensions*: Ian P. Howard and Brian J. Rogers, *Perceiving in Depth, Volume 2: Stereoscopic Vision*. New York: Oxford University Press, 2012.

20 *Sir Charles Wheatstone first identified stereopsis*: Charles Wheatstone, F.R.S., Professor of Experimental Philosophy in King's College, London, *Contributions to the Physiology of Vision.—Part the First. On some remarkable, and hitherto unobserved, Phenomena of Binocular Vision.* http://www.stereoscopy.com/library/wheatstone-paper1838.html.

22 *Over 250,000 image cards (or stereographs)*: Barbara Stafford, Frances Terpak, and Isotta Poggi, *Devices of Wonder: From the World in a Box to Images on a Screen*. Los Angeles: Getty Research Institute, 2001.

22 *sold around half a million viewers*: Martin Kemp, *The Science of Art: Optical Themes in Western Art from Brunelleschi to Seurat*. New Haven: Yale University Press, 216.

23 *"I felt sure this was decidedly better"*: Oliver Wendell Holmes, "The American Stereoscope," *The Philadelphia Photographer*, January 1869. Reprinted in *IMAGE, Journal of Photography of the George Eastman House*, vol. 1, no. 3 (March 1952).

24 *"Form is henceforth divorced"*: Oliver Wendell Holmes, "The Stereoscope and the Stereograph," *The Atlantic*, June 1859. https://www.theatlantic.com/magazine/archive/1859/06/the-stereoscope-and-the-stereograph/303361/.

25 *H. C. White advertised the model*: "A Short but Accurate History of H.C. White Stereoscopes and Stereoviews—in the Company's Own Words—Published In 1908." https://www.flickr.com/photos/okinawa-soba/5070181314/sizes/l/. Accessed January 28, 2017.

31 *"Five-cent theatres abound"*: "A Theater with a 5,000,000 Audience," in *The World's Work: A History of Our Time*, Volume XX (Garden City, New York: Doubleday, Page & Company, May 1910), 12,876.

32 *According to a review in the* **New York Dramatic Mirror**: *New York Dramatic Mirror*, June 16, 1915, quoted in Ray Zone, *Stereoscopic Cinema and the Origins of 3-D Film, 1838–1952*. Lexington: University Press of Kentucky, 2007.

35 *"Seeking a novelty to charm its fickle audiences"*: Frank S. Nugent, "Meet the Audioscopiks. Metro's New Three-Dimensional Film Is

the Next Novelty on the Schedule," *The New York Times*, December 8, 1935.

38 **By 1950, about one out of every ten American households:** "Facts-Stats." http://www.tvhistory.tv/facts-stats.htm. Accessed March 30, 2017.

38 **An article in the December 15, 1952, edition of Life:** "Miscellany: An Eyeful at the Movies," *Life*, December 15, 1952, 146.

40 **"The result has a kind of spellbinding effect":** *The Hollywood Reporter*, as reported in www.3dfilmarchive.com/home/widescreen-documen tation. Accessed April 1, 2017.

40 **"What does the future hold for 3-D?":** John Norling, "3-D Movies . . . Epitaph or Prologue?" *International Projectionist*, July 1954. As quoted on http://www.3dfilmarchive.com/what-killed-3D. Accessed April 1, 2017.

CHAPTER 2—THE ULTIMATE DISPLAY

41 **"a photographic view of the scene as a human pair of eyes would see it":** Richard F. Dempewolff, "Movies on a Curved Screen Wrap You in Action," *Popular Mechanics*, August 1952.

43 **"Every capable artist has been able":** Morton Heilig, "The Cinema of the Future," 1955. https://gametechdms.files.wordpress.com/2014/08/w6_thecinemaoffuture_morton.pdf. Originally published in Spanish as "El cine del futuro," *Espacios* no. 23–24 (January–June 1955).

43 **"in all its magnificent colors":** Ibid.

44 **"I sat down, put my eyes and ears in the right places":** Howard Rheingold, *Virtual Reality: The Revolutionary Technology of Computer-Generated Artificial Worlds—and How It Promises to Transform Society.* New York: Simon & Schuster, 1992, 50.

45 **After the motorcycle ride:** Ibid., 55.

45 **"As a result of this situation":** Morton Heilig, Senorama Simulator, US Patent #3,050,870. https://patents.google.com/patent/US3050870, August 28, 1962.

47 **"A display connected to a digital computer":** Ivan E. Sutherland, "The Ultimate Display." *Information Processing 1965: Proceedings of IFIP Congress 65*, vol. 1. London: Macmillan and Company, 1965, 506–508.

48 **"The ultimate display would":** Ibid.

49 **"Even with this relatively crude system":** Ivan E. Sutherland, "A Head-Mounted Three-Dimensional Display." *Proceedings of the AFIPS Fall Joint Computer Conference.* Washington, DC: Thompson Books, 1968, 757–64.

49 **His teachers would let him tinker:** Fred Moody, *Visionary Position: The Inside Story of the Digital Dreamers Who Are Making Virtual Reality a Reality.* New York: Crown Business, 1999, 11.

CHAPTER 3—CONSOLE COWBOYS

55 *"live as obscurely as possible"*: Jaron Lanier, *Dawn of the New Everything: Encounters with Reality and Virtual Reality.* New York: Holt, 2017, 9.

55 *"oddballs"*: Ibid., 18.

56 *"The overall form reminded me"*: Ibid., 28.

56 *"Reading about Ivan's work"*: Ibid., 45–46.

58 *"I said, 'Oh, VPL'"*: Burr Snider, "Jaron." *Wired,* February 1, 1993. https://www.wired.com/1993/02/jaron/.

59 *"All of a sudden we had a company"*: Ibid.

59 *"Potential investors would come around"*: Ibid.

59 *hand wearing the high-tech gadget: Scientific American,* October 1987.

60 *"doing nothing less than inventing"*: Steven L. Thompson, "In a Super Cockpit," *Washington Post,* May 10, 1987.

61 *"The most remarkable thing was the phone calls"*: Thomas A. Furness III, "Being the Future." Lecture at Augmented World Expo, Santa Clara, CA, June 10, 2015.

62 *"I realized we needed to get this technology out in the world"*: Ibid.

62 *"may someday send the world"*: Richard Scheinin, "Through the Looking Glass," *Chicago Tribune,* February 18, 1990.

62 *"we are witnessing the birth of Virtual Reality"*: David Gale, "At the Other End of Nature," *The Guardian,* April 13, 1990.

63 *"babe in the woods"*: Lanier, *Dawn of the New Everything,* 187.

63 *"One time I signed"*: Ibid.

65 *John O'Neill told the Associated Press*: Frank Baker, "Hasbro Abandons Game After Spending $59 Million on Research," Associated Press, July 19, 1995.

CHAPTER 4—INTO THE RIFT

76 *"I felt like I was literally"*: Larry Frum, "My 5 Favorite Highlights from E3, CNN.com, June 7, 2012. https://www.cnn.com/2012/06/07/tech/gaming-gadgets/e3-highlights/index.html." Accessed May 7, 2017.

80 *bare-bones room with a short set of stairs*: Jerry Beilinson, "Palmer Luckey and the Virtual Reality Resurrection," *Popular Mechanics,* May 28, 2014. https://www.popularmechanics.com/technology/gadgets/a12956/palmer-luckey-and-the-virtual-reality-resurrection-16834760/. Accessed May 8, 2017.

CHAPTER 5—TWO BILLION REASONS

86 *"virtual reality's greatest hope"*: Will Greenwald, "CES 2013: Hands On with the Oculus VR Rift, Virtual Reality's Greatest Hope," PC

Magazine.com, January 9, 2013. https://www.pcmag.com/article2 /0,2817,2414070,00.asp.

95 *"I feel like I've been waiting in line"*: Amoliski, reply to "Facebook Acquires Oculus VR," Reddit, March 25, 2014, https://www.reddit.com/r/ oculus/comments/21cvry/facebook_acquires_oculus_vr/cgbtzzf.

95 *"I don't want a social platform"*: Christina Warren, "Facebook Acquires Oculus VR for $2 Billion," Mashable, March 25, 2014. https:// mashable.com/2014/03/25/facebook-acquires-oculus-vr-for-2-billion/ #0.pcj4VPkmqg; https://www.reddit.com/r/virtualreality/comments/ 21d19k/facebook_acquires_oculus_vr_for_2_billion/cgc0mqr/.

95 *"Facebook is not a company of grass-roots tech enthusiasts"*: Markus Persson, "Virtual Reality Is Going to Change the World," March 25, 2014. http://notch.net/2014/03/virtual-reality-is-going-to-change-the -world.

102 *"an early-access, beta-version"*: "Introducing the Samsung Gear VR Innovator Edition," Oculus blog, September 3, 2014. https://www.oculus .com/blog/introducing-the-samsung-gear-vr-innovator-edition/.

102 *"estimated 19 million"*: Shane Schick, "Evans Data: Mobile Developers Now Number 8.7 Million Worldwide," Fierce Wireless, June 20, 2014. http://www.fiercewireless.com/developer/evans-data-mobile -developers-now-number-8-7-million-worldwide.

103 *"breathtaking"*: Peter Rubin, "Oculus' Mind-Blowing New Prototype Is a Huge Step Toward Consumer VR," *Wired,* September 22, 2014. https:// www.wired.com/2014/09/oculus-crescent-bay-prototype/.

103 *"a new high-water mark in virtual reality"*: Kyle Orland, "Oculus' Crescent Bay Prototype Is a New High-Water Mark," *Ars Technica*, September 22, 2014. https://arstechnica.com/gaming/2014/09/eyes-on-oculus -crescent-bay-prototype-is-a-new-high-water-mark/.

CHAPTER 6—TAKING HOLD

117 *The device, called Google Cardboard*: Frederic Lardinois, "The Story Behind Google's Cardboard Project," TechCrunch, June 26, 2014. https://techcrunch.com/2014/06/26/the-story-behind-googles -cardboard-project/. Accessed July 30, 2017.

118 *"We believe that virtual reality"*: Natasha Lomas, "HTC and Valve Partner to Make a VR Gaming Headset," TechCrunch, March 1, 2015. https://techcrunch.com/2015/03/01/htc-vive/.

121 *"absolutely incredible"*: Paul James, "30 Minutes Inside Valve's Prototype Headset: Owlchemy Labs Share Their Steam Dev Days Experiences," RoadtoVR, January 30, 2014. https://www.roadtovr.com/hands

-valves-virtual-reality-hmd-owlchemy-labs-share-steam-dev-days
-experiences/.

121 *"world changing"*: Fraser Brown, "Valve's VR Demo Impresses at steam
Dev Days," PCGamesN, December 2013. https://www.pcgamesn.com/
valves-vr-demo-impresses-steam-dev-days.

121 *"lightyears ahead of the original"*: Dave Cook, "Valve VR Prototype Is
'Lightyears Ahead of the Oculus Dev Kit,' Says Dev After Studio Visit,"
VG24/7, February 28, 2014. https://www.vg247.com/2014/02/28/valve
-vr-prototype-is-lightyears-ahead-of-the-original-oculus-dev-kit
-says-dev-after-studio-visit/.

CHAPTER 7—VR AND CODING IN LAS VEGAS

141 *"virtual reality is finally coming to the living room"*: Adam Rogers,
"Through a Glass, Darkly," *Newsweek*, January 25, 1995.

143 *"a very, very big technological breakthrough"*: "HTC Vive To Demo A
'Very Big' Breakthrough In VR At CES," Engadget, December 18, 2015.
https://www.engadget.com/2015/12/18/htc-vive-vr-big-breakthrough
-ces/.

143 *"We shouldn't make our users swap their systems"*: Ibid.

153 *"the Oculus Rift is the most revolutionary gaming experience"*: Verge
Staff, "The Verge Awards: The Best Of CES 2013," The Verge, January
11, 2013. https://www.theverge.com/2013/1/11/3865786/verge-awards
-ces-2013.

153 *"the coolest thing we've ever put on our face"*: *Wired* Staff, "The Best
Of CES 2013," *Wired*, January 13, 2016. https://www.wired.com/2013
/01/the-best-of-ces-2013/.

163 *planned to start installing the demo*: Nathan Ingraham, "Audi Is Outfit-
ting Its Dealers with an Impressive VR Experience," Engadget, January
10, 2016. https://www.engadget.com/2016/01/10/audi-vr-dealership
-car-configurator/.

CHAPTER 8—THE ONCOMING TRAIN

170 *I didn't spend all those years playing Dungeons & Dragons*: "Jose
Chung's from Outer Space," twentieth episode of *The X-Files*. Originally aired April 12, 1996 on FOX.

CHAPTER 9—THIS IS REAL

189 *"I got an e-mail that was very nondescript"*: Ben Kuchera, "Oculus
Founder: 'I'll Be Damned If Some Random Delivery Guy Will Deliver
the First Rift,'" Polygon, March 26, 2016. https://www.polygon.com/2016
/3/26/11310190/oculus-delivers-rift-luckey. Accessed December 4, 2017.

189 *"This is incredible"*: Palmer Freeman Luckey, "Personally delivering the first Rift to Alaska!" Facebook, March 26, 2016. https://www.face book.com/palmer.luckey/videos/vb.1063830478/10207710972306676/.

190 *"I was pretty adamant"*: Kuchera, "Oculus Founder: 'I'll Be Damned . . .'"

190 *"Hiking through Alaska"*: Palmer Luckey, Twitter, March 26, 2016. https://twitter.com/palmerluckey/status/713790929836965888.

199 *"Every core feature of both the Rift and Vive"*: "Oculus Becoming Bad for VR Industry?" Reddit thread, 2017. https://www.reddit.com/r/Vive/comments/4klu94/oculus_becoming_bad_for_vr_industry/d3g6.

203 *SuperData Research estimated Sony would sell*: Ben Reeves, "Playstation VR Estimated To Sell 2.6 Million Units By End Of Year," Game Informer, October 12, 2016. http://www.gameinformer.com/b/news/archive/2016/10/12/playstation-vr-estimated-to-sell-2-6-million-units-by-end-of-year.aspx.

CHAPTER 11—MAGICAL THINKING

226 *children's book author L. Frank Baum*: L. Frank Baum, *The Master Key: An Electrical Fairy Tale*. Indianapolis, IN: Bowen-Merrill Company, 1901, p. 94. https://archive.org/details/masterkeyanelec00corygoog.

228 *"A well-designed virtual reality"*: Steven Feiner, Blair MacIntyre, and Dorée Seligmann, "Knowledge-Based Augmented Reality." *Communications of the ACM*, vol. 36, no. 7 (July 1993): 53–62.

228 *"We have developed a preliminary set"*: Ibid.

229 *Louis Rosenberg, a Stanford University graduate student*: Louis B. Rosenberg, "The Use of Virtual Fixtures to Enhance Operator Performance in Time Delayed Teleoperation," Defense Technical Information Center, March 1993. http://www.dtic.mil/dtic/tr/fulltext/u2/a296363.pdf.

230 *worst innovations in sports history*: Page 2 staff, "Worst Sports Innovations," ESPN Page 2. http://www.espn.com/page2/s/list/innovations/worst.html. Accessed October 20, 2016.

EPILOGUE—THE AGE OF THE UNREAL

246 *the AR market could approach 3.5 billion installed devices*: "Ubiquitous $90 Billion AR To Dominate Focused $15 Billion VR By 2022," Digi-Capital blog, January 26, 2018. https://www.digi-capital.com/news/2018/01/ubiquitous-90-billion-ar-to-dominate-focused-15-billion-vr-by-2022/.

Index